ECHO
OF THE
BIG
BANG

ECHO
OF THE
BIG
BANG

With a new epilogue by the author

Michael D. Lemonick

PRINCETON UNIVERSITY PRESS
PRINCETON AND OXFORD

Third printing, and first paperback printing,
with a new epilogue by the author, 2005
Paperback ISBN 0-691-12242-3

The Library of Congress has cataloged the cloth edition
of this book as follows

Lemonick, Michael D., 1953–
Echo of the big bang / Michael D. Lemonick.
p.   cm.
Includes index.
ISBN 0-691-10278-3 (alk. paper)
1. Microwave Anisotropy Probe (Spacecraft) 2. Space probes—Design
and construction. 3. Big bang theory. 4. Princeton University—Research.
5. Goddard Space Flight Center—Research. I. Title.
TL795.3 .L45   2003
523.1—dc21   2002042721

British Library Cataloging-in-Publication Data is available

This book has been composed in Sabon

Printed on acid-free paper. ∞

pup.princeton.edu

Printed in the United States of America

5   7   9   10   8   6   4

For Hannah

Contents

CHAPTER 1

Is Something Amiss
in the Universe?

Spring does not come subtly to Princeton University. It's early April 2002, and even after a mild winter, the brilliance of the campus in full bloom is almost overwhelming. The air is still uncomfortably cool, but vivid color is visible no matter where you look. Daffodils and hyacinths cover the ground; forsythia bushes burst with yellow flowers at eye level; cherry and magnolia trees are clouds of pink and white overhead. David Spergel doesn't notice any of this. He's much too distracted. A few weeks ago, he discovered evidence of something surprising and unsettling about the universe, evidence suggesting that its fundamental character is not what astrophysicists have believed for the past two decades or so. Papers published by other astrophysicists over recent months have declared that the cosmos is, at last, well understood. Modern cosmology, they say—the branch of astronomy that dares to address the ultimate questions about the birth and death of the universe—is essentially solved, only eighty years or so after it was founded. All that remains is to tidy up the last few decimal points. Newspapers and magazines have dutifully reported this comfortable conclusion, much as they reported a strangely similar comment made by the American physicist Albert Michelson in 1894: "While it is never safe to say that the future of Physical Science has no marvels even more astonishing than those of the past, it seems probable that most of the grand underlying principles have been firmly established and that further advances are to be sought chiefly in the rigorous application of these principles to all the phenomena which come under our notice." Within a few years after that declaration, physicists would discover such previously unsuspected phenomena as radioactivity, subatomic particles, relativity, and quantum mechanics.

But David Spergel has reason to believe that cosmology is not solved. For a few hours he was the only one on Earth who had evidence to support this doubt. He may even have been the only one in the universe who had it. Now a small group of colleagues, fewer than twenty in all, have seen the evidence as well. They'll tell the rest of the world early the following year, most likely at a press conference at NASA headquarters. Their announcement will almost certainly be accompanied by the sort of public-relations blitz NASA has perfected in more than forty years of space exploration and discovery. The space agency puts on press conferences at different levels of breathlessness, from mild to hyperventilation, depending on what's being announced. The discovery of a new isotope of tin in the dust that floats between the stars doesn't generate much enthusiasm. The claim of evidence of life in a Martian meteorite gets the full treatment.

But for now, Spergel is focused on convincing himself and the others that what he's found is real rather than some glitch in the satellite that has been scanning deep space for the past nine months, or a bug in the computer code he uses to analyze its observations. Either of these is possible, for no matter how careful Spergel and his colleagues at Princeton and a handful of other institutions have been, the chance of a mechanical or electronic or software flaw, either new or too subtle to have been noticed over the past six years of careful work, may be making its presence known now, at the worst possible moment. The last thing these astronomers, physicists, and engineers want to do is make the shocking claim that much of the work of cosmology over the past two decades has been based on a faulty theoretical foundation—and then, a few months later to say, "Never mind."

This is looking increasingly unlikely, though. At first, when he told the others about what he'd found, they were appropriately skeptical. Spergel is one of the smartest, most talented theoretical astrophysicists around. That's why he was recruited for this project in the first place, and why, in the fall of 2001, the forty-year-old scientist had won a MacArthur Foundation "genius" grant. But even geniuses can make a mistake when they're writing hundreds of thousands of lines of computer code. Even geniuses can think they see patterns in data when the patterns aren't really there. Even geniuses are human enough to leap to world-shaking conclusions while overlooking a mundane explanation for what they've evidently found. Besides, the others on the team—all of them extraordinarily bright, even if the MacArthur committee hasn't formally certified them as such—have also looked at the data, and they can't think of any mundane explanation, either. The computer code is working fine. The satellite is performing as close to perfectly as anyone could wish. And it's simply not telling the story everyone expected.

In principle, of course, scientists aren't supposed to expect anything when they go into an experiment. They're simply supposed to observe, as objectively as if they were robots, aware of, but unprejudiced by, what their predecessors have seen. But they aren't robots. Science is an intensely human enterprise. Observers go into a new experiment with both expectation and hope. Unless it's the first time an experiment is being done, they generally expect that existing theories or assumptions are probably correct, that this latest attempt to observe and measure will expand or refine what we already know. This is true often enough that it can prove dangerous: people tend to see what they're expecting to see. When that happens, says Tod Lauer, an astronomer at Kitt Peak National Observatory in Arizona, "you're not likely to double-check it very carefully. Whereas if you see something unexpected, you recheck it over and over to figure out where you might have been fooling yourself." Yet the "correct" observation could be equally wrong. An observer can be too quick, in other words, to dismiss a crucial, telltale anomaly as nothing more than experimental error.

Scientists can also be tempted to err in precisely the opposite direction. It's important and satisfying to confirm or refine an existing theory with higher precision than anyone's ever achieved before. But most scientists agree that it's even more satisfying to prove the conventional wisdom utterly wrong. Showing yet again that Einstein's general theory of relativity is correct is certainly a good thing; refuting it would be a much bigger deal (that's one reason Einstein refutations are the number-one choice of cranks who send their handwritten "manuscripts" to physicists and science journalists).

So if a scientist discovers something dramatically new and important—cold fusion, say, or the first evidence of a planet orbiting a star other than the sun—it can be tempting to shout the news before you've thought it through. In the former case, two chemists from the University of Utah coined a new shorthand for "discovery that really isn't." But even when you have thought it through, you can overlook something. In 1991, a radio astronomer named Andrew Lyne thought he'd discovered planets orbiting a pulsar, the dense, burned-out remnant of an exploded star. He knew this was an audacious, even a preposterous claim, so he did every test he could think of to explain it away as a glitch. Eventually, he went public, only to realize to his horror that he'd failed to think of the one test that actually could—and in fact did—prove that the planets weren't there after all. Lyne's public apology to the astronomical community was deemed by John Bahcall, then president of the American Astronomical Society, to be "the most honorable act I've ever witnessed." But that didn't make Lyne feel much better.

Finally, there's a more subtle source of confusion in presenting new data. It's common that the first studies or experiments to explore a scientific question are inconclusive, but suggestive. The instruments in question—the telescopes, or particle detectors, or seismographs—are pushed to their limits of sensitivity, and find evidence that's not quite definitive. Depending on their confidence, researchers might play up or play down what they've found—label it a tentative discovery, or merely an interesting result. A long-sought particle known as the Higgs boson may have turned up in 2000, for example, in experiments at the Large Electron-Positron Collider in Europe. Or it may not; because of a scheduled major upgrade of the equipment, the experiment couldn't be run long enough to make a definitive measurement— although the physicists pleaded for a few more months. These scientists opted for caution, and didn't claim a discovery.

In medicine, by contrast, the public demands to hear about every result, preliminary or not, and often acts on it. When doctors found a relationship between fresh vegetables and reduced risk of cancer, they deduced that beta carotene, a chemical found in many vegetables, was a likely reason. Beta carotene supplements, they said, might be a good idea. So people began swallowing beta carotene pills by the handful. Later studies showed that taking these supplements actually raised the risk of cancer in some people. Something similar happened when doctors first established a link between saturated fats and heart disease. They suggested it might be wise to switch from butter to margarine. Then, a few years later, they discovered that the processed, or hydrogenated, vegetable oil in margarine was actually worse for the heart than ordinary saturated fat. So the recommendation swung back. People were indignant and assumed that medical scientists didn't know what they were talking about. But the earlier recommendations were as good as the data permitted them to be. The mistake, largely the fault of health experts and journalists, was that the provisional nature of the research was downplayed in the interest of making a good story.

Among the sciences, cosmology is especially prone to the danger of premature conclusions. One reason is that it's not an experimental science. There is only one universe, and it's physically inaccessible. You can't deduce its underlying structure or behavior or laws by taking one apart in the lab, or by varying the growth medium and cultivating a new one to see what happens. It's hard, moreover, to gather information about the cosmos; the photons of electromagnetic radiation that carry information about the stars and galaxies are sparse, and they overlap with each other in a tangle of data that must be untangled. As a result, astronomers have always been forced to build their models of the universe, initially at least, on meager information. A century ago it wasn't even clear that a universe existed beyond the Milky

Way. Eighty years ago, nobody imagined that the universe was expanding. Forty years ago, the Big Bang was a somewhat crackpot theory.

Time after time, astronomers have been startled to realize how much less they understood about the universe than they'd thought. Confident statements about its basic nature have been proven not just wrong, but deeply, profoundly and sometimes embarrassingly wrong. In almost every case, the mistakes have been based on incomplete information, which in turn has been the fault not of sloppy observers but of primitive technology. In the 1920s, when the modern picture of the cosmos began to emerge, the largest and most powerful telescope on Earth had a light-gathering mirror just 100 inches across; the largest today spans nearly 400. The most sensitive medium for recording that light was the photographic plate, which was much better than the human eye but still very inefficient and inconsistent; today's charge-coupled devices, or CCDs, are a hundred times better.

And that just covers visible light. Nobody suspected a century ago that it would be useful to explore the full range of the electromagnetic spectrum, of which visible light is only a small subset. As the wavelength of light becomes shorter and shorter, visible light shades into ultraviolet, which humans can't detect (but which causes sunburn nonetheless). As the wavelength shrinks further, ultraviolet gives way to X-rays, then gamma rays. All three are part of the electromagnetic concerto broadcast by the universe, by individual stars and black holes and knots of superheated gases; all three carry telltale messages about the nature of these phenomena; yet none of them was part of astronomers' observing programs, nor even contemplated.

The same applied to the region of the spectrum with wavelengths longer than those of visible light. If short-wavelength radiation corresponds to the right-hand keys, the high-pitched notes, on a cosmic piano, then infrared radiation, microwaves, and radio waves are the bass notes to the left. By the 1920s, physicists understood something about radio waves, and broadcasters had begun to exploit them, first to send Morse code across the oceans and then to send news, and music, and jingles for selling detergent. In the early part of the twentieth century, Guglielmo Marconi even aimed his primitive wireless antenna toward the planet Mars, to try and pick up any broadcasts the Martians might be directing toward their sister planet.

But Marconi never tried to listen for natural radio emissions from the heavens. Nobody did. And so the deepest tones of the cosmic electromagnetic concerto, like the high notes, went unheard. Astronomers tuned in only to the few notes they could get to at the center of the keyboard. What they learned was true, as far as it went. But the bass and treble notes they couldn't hear would, time and again, alter the melody beyond recognition. Quasars, black holes, neutron stars, dark clouds of organic chemicals between the stars were all invisible, and in many cases unimagined, until astronomers learned to probe the high- and low-pitched frequencies of light.

So it was as well with the experiment David Spergel and his colleagues are now engaged in. Anyone who watched television before the days of cable, or who still gets a TV signal out of the air, remembers "snow"—the salt-and-peppery visual static that filled the screen when a set was tuned to a weak or nonbroadcasting channel. Nobody ever gave it much thought, except to curse at it; those who did figured it was some electronic noise in the picture tube, or maybe a distant station, coming through so feebly that the picture had disintegrated.

But at least some of those electronic crackles were and are something much more important than that. They're a message from the birth of the universe—a detailed record of the beginnings of space and time, and of the subsequent evolution of the cosmos. Every minute of every day, the Earth is bombarded with a barrage of photons, the particle-like building blocks of electromagnetic radiation. Most of these come from the Sun and the stars; they were emitted anywhere from today to a few thousand years ago. The photons that help wash out the *Today Show* and *Sesame Street*, though, are thousands of times older. They are by far the oldest radiation in the universe—the electromagnetic echo of the Big Bang itself. These photons began their journey through intergalactic space about 14 billion years ago. At that time, they carried an intensity as bright as that of the Sun. If we'd been around to see them, we'd have been blinded by a yellow-white brilliance bombarding us from all directions at once. Now, enfeebled during their long journey, they've cooled from about 6000° Celsius to −270°—a bit less than three degrees above absolute zero, the coldest temperature possible. Their nature has shifted too, down from visible light to near infrared and finally into the microwave part of the spectrum—a bass note in the electromagnetic keyboard, not quite as deep as a radio wave, but much lower than the middle octave of ordinary light.

Ever since Einstein founded the science of theoretical cosmology by treating the universe of space and time as a distinct object with distinct attributes, and especially since Edwin Hubble discovered the expanding universe in 1927, physicists and astronomers have tried to divine the origin and fate of the universe. Stars, planets, and even galaxies come and go, but they do so on a grand stage that has its own, independent history. But it wasn't until the 1940s that theorists realized that the expanding universe must have loosed a burst of hot photons, and it wasn't until 1965—an astonishingly long interlude, in retrospect—that observers first identified them. When John F. Kennedy was assassinated—an event that seems to tens of millions of Americans as though it happened yesterday—the Big Bang model of the universe was still considered a long shot, and rightly so.

The fact that this background glow of cosmic microwaves existed at all was powerful evidence that the Big Bang had taken place—an unmistakable announcement that the universe had a birth date. But theorists quickly realized that it could also be used as a powerful diagnostic tool. The modern universe, they knew, is lumpy—mostly empty space, punctuated, at varying levels of organization, by stars, galaxies, clusters, and superclusters of galaxies. These lumps must have started as variations in density in the newborn universe, which grew, under gravity, into their present form. But since the newly found glow of microwaves was emitted when the universe was a mere 300,000 or so years old, any density variations present at the time should have left their imprint on it. A slightly overdense region would have been very slightly hotter than average, while a slightly underdense region would have been cooler. And these temperature differences should still be detectable, even after 14 billion years.

The hot and cold spots, moreover, should be a direct consequence of the physical conditions present at the time—of the temperature and pressure and composition and previous history of the young universe. If you could see them sharply enough, you could figure out how big they were, and how much hotter the hot was than the cold, and the ratio of big hot spots to medium to small to tiny. And using standard physics, that information would in turn tell you what the young universe was made of (a bowl of Jell-O vibrates differently from a bowl of oatmeal, after all), and how dense it was, and what forces were in play at the time. By comparing maps of the microwave background radiation with maps of the present-day cosmos, astronomers could also piece together the story of how we got from there to here—exactly how the galaxies formed, and how the Earth and its inhabitants ultimately came to be.

With all this information hidden within it, the cosmic microwave background radiation, or CMB, was the astronomical equivalent of the human genome. Just as the genome bears all of the data required to manufacture and operate a human being, the microwave background encodes all of the information—all the initial conditions and physical laws—for making and operating a universe.

The major difference is that the genome has been relatively simple, if laborious, to read, but so complex that geneticists don't expect to understand it fully for decades. The genome of the universe, by contrast, would be simple to understand if only astrophysicists could read it. It's so faint, though, and so badly contaminated with other sorts of radiation from our own planet and the solar system and the galaxy that only in the 1990s could astronomers finally see the hot and cold spots for the first time. It took a satellite, the Cosmic Background Explorer (COBE), to do so. Seeing anything at all after so many years of searching was so exciting that astrophysicist George Smoot, the principal investigator on one of COBE's key instruments, declared at the time that "it's like seeing God." He could have said more accurately that it's like seeing God through Coke-bottle eyeglasses that haven't been cleaned for a year. The satellite showed for the first time that the spots were there, which came as a relief to cosmologists who were beginning to wonder. But the images were much too crude to say much more than that.

Still, COBE did help solidify the so-called Standard Model of the universe, a model that had begun taking shape in the 1920s with the discovery of the expanding universe. Over the years it had been modified to accommodate a number of theories and discoveries—that the early universe was hot and dense, that the cosmos is suffused with mysterious "dark matter," probably in the form of a yet-undiscovered particle; that spacetime underwent a brief but dizzying period of hyperexpansion known as inflation. But just as was the case forty or fifty or eighty years ago, suggesting that these elements of the Standard Model are "known" is an intellectually dangerous concept. Now, as then, some of what's known is almost certainly wrong, perhaps subtly, perhaps egregiously. The best way to put the Standard Model on a firm footing, or, alternatively, to expose its unsuspected weaknesses, is to take a harder, sharper look at the CMB; to see the details the COBE satellite couldn't; to read the genome of the universe with high precision.

In the aftermath of the COBE satellite, that's just what cosmologists proposed to do. They would build another satellite to take the next step. This time, they wouldn't be satisfied just with detecting the cosmic ripples: they'd also measure their intensity and distribution and characteristic sizes. They'd try do it faster, and for less money than COBE had used up; Daniel Goldin, the NASA administrator, proclaimed in the early 1990s that henceforth his agency would be doing everything "better, faster, cheaper." COBE, which had cost $500 million and taken an agonizing fifteen years from concept to launch, was a good illustration of what he wanted not to repeat. Even under the best of circumstances, though, it would take years to decide precisely what the design of the new satellite should be and who would build it, and then to construct and launch such a complex, delicate piece of machinery.

One team of cosmologists would be selected among several applicants to design and build the new satellite, which would ultimately be named MAP, the Microwave Anisotropy Probe. But others would try to make the measurements from the ground or from balloons floating in the stratosphere. Their measurements would be contaminated with microwave emissions from the air and the ground, and would thus be harder to decipher. They would be able to scan only a small part of the sky. As with a public opinion poll, it might be possible to get a good idea of the overall situation by surveying only a fraction of the available data. But it also might be highly misleading; if you'd polled voters only in New York City prior to the 2000 presidential election, you'd have predicted a landslide for Al Gore.

These limited experiments might, in other words, answer some of the great open questions in cosmology before the satellite could, but it would be hard to know for sure that they'd really done it. Sometimes the marginal experiments that precede a definitive one are wrong or misleading. Sometimes they're right but are ignored because someone failed to make a crucial connection between theory and observation, or because the field is not yet mature enough to put things into their proper context. The expanding universe was officially discovered in 1927, but the evidence was already there in 1920. The CMB was arguably first detected in 1939, a quarter-century before the formal discovery, but it went unrecognized. The experiment that was finally credited with seeing the CMB in 1965 had first taken place in 1961 and been analyzed—incorrectly—in 1962. The scientists who did it decided there was nothing there. Only when a second team of observers tried again did they make the discovery. And while COBE's detection of ripples in the CMB made headlines, the ripples were actually seen a few months earlier by microwave detectors in several other experiments—but the physicists who saw them weren't confident enough of their results to report them.

Mindful that they might be able to find at least some of the answers MAP would be seeking before the satellite could be launched, ground-based observers would be working harder than ever through most of the 1990s. Given the drawbacks of their experiments, they would have to weigh carefully just how early and how definitively to present whatever results they might get; it would be professionally delicious to scoop MAP, but it would be embarrassing to be first but wrong. Nevertheless, it would be easier and quicker to take measurements from the ground than from space. In the decade that would ultimately elapse between COBE and MAP, plenty of other groups would make discoveries about the cosmic microwave background. Some might say "margarine" when the real answer would turn out to be "butter."

But MAP, when it finally reported in, would end much of the debate. When great questions about the universe are first asked, the relevant observations are almost always too sparse and too unreliable to give solid answers. The data eventually become good enough to change that, the questions are laid to rest—and sometimes, unsuspected questions are newly posed. Thanks to its sensitivity and extraordinary performance, David Spergel knows that MAP will do the former, and he hopes for the latter as well. He just needs to make sure that he understands what MAP is telling him before he tells it to the world.

CHAPTER 2

The Birth of Cosmology

On April 26, 1920, David Spergel's revelation about the nature of the universe lay eight-two years in the future. Nobody had begun to wrestle with the questions cosmologists would be arguing over fourscore years hence. Nobody would even have comprehended them, probably, and if a time traveler could somehow have come back and tried to explain what the COBE satellite did and what MAP is intending to do, he or she would probably have been dismissed as an utter lunatic. The notion of a beginning to time and space was something even H. G. Wells, with his vivid and frequently prescient imagination, hadn't addressed.

Yet in a broad sense, the issue on the table as members of the National Academy of Sciences gathered for a discussion that evening was precisely similar to those of today: it addressed a crucial open question about the nature of the universe for which nobody had a convincing answer. At 8:15, Harlow Shapley, of the Mount Wilson Observatory in the mountains near Los Angeles, and Heber Curtis, of Lick Observatory near

San Jose, would mount the stage and debate "The Scale of the Universe." Shapley's position was that it was large; Curtis argued that it was quite small—but as the historian of astronomy Michael Hoskin has shown, the meat of the debate actually took place in two papers published after the debate, not during the event itself. While these two eminent observers did give their opposing views on how big the Milky Way galaxy is, moreover, and where the Sun is located within it, the debate is much better remembered for what they had to say on a related topic: Is the Milky Way galaxy in which the Sun and Earth are embedded the entire universe? Or do other, similar galaxies lie beyond its edges, thus making the Milky Way one small element in a huge cosmos?

Originally, the National Academy had considered giving the evening over to a discussion of the new theory of general relativity proposed by a Professor Einstein instead. But that theory, published four years earlier and supported by a dramatic observation just a year or so before the debate, was considered a bit too trendy. Hoskins quotes Mr. Abbot, secretary of the Academy, as writing: "I pray to God that the progress of science will send relativity to some region of space beyond the fourth dimension, from whence it may never return to plague us."

The key question for both Shapley and Curtis was the so-called spiral nebulae—tiny patches of fuzzy light, visible in most cases only through telescopes (one exception is the Great Nebula in Andromeda, whose core, though not its spiral arms, can be seen if you can get far enough from cities and suburbs). Observers had known since the 1850s that some of these blobs had distinctly spiral structures, and Shapley was convinced that they were clouds of glowing gas within the Milky Way. His chief argument was his own conviction that the Milky Way was a gigantic collection of stars some 250,000 light-years across. If our home galaxy were that big, the spiral nebulae would have to be impossibly distant to lie far beyond its edge. Besides, the nebulae were blurs of light; if they were made of stars, you should be able to see those stars, just as you could resolve the blur of the Milky Way itself into stars. And finally, the respected Dutch astronomer Adriaan van Maanen had observed that some of the spirals had measurable spins. Unless they were nearby, these rotations would be absurdly fast.

But Curtis was sure the Milky Way was only about 30,000 light-years across. The spiral nebulae could therefore be distinct galaxies—"island universes," in the words of the philosopher Immanuel Kant in the 1700s, who had proposed their existence before the spirals were even discovered. Curtis had other lines of argument as well. One was based on analysis of the nebulae's light. Astronomers had already known for decades that when you smear the light of a star into its constituent colors—just as raindrops turn sunlight into a rainbow—you can, with a device called a spectrograph, see thin black lines interrupting the spectrum of light. They're caused by atoms in the star's atmosphere that absorb specific wavelengths of light, and by their positions on the spectrum they can tell you what chemical elements a star is made of. Many of the nonspiral nebulae had spectra that looked nothing like those of a star; these were almost certainly clouds of gas. But the spectra of spiral nebulae were impressively starlike. Thus, he reasoned, they must be collections of stars so distant that they merged into an unresolvable blur. Curtis also argued that you *could* see individual stars in the spiral nebulae: once in a while, a single spot of light would blaze within the general glow. These, he said, were novas, just like the flaring stars occasionally seen in the Milky Way.

In the end, Curtis turned out to be right about the spiral nebulae—but some of his arguments were invalid. Shapley reached the wrong conclusion, based in part on incorrect observations. And that, argues astronomer Frank Shu in his 1992 textbook, *The Physical Universe: An Introduction to Astronomy*, makes the debate a useful lesson for modern astronomers. "The Shapley-Curtis debate was . . . important," he writes, "not only as a historical document, but also as a glimpse into the reasoning processes of eminent scientists engaged in a great controversy for which the evidence on both sides is fragmentary and partly faulty." The rotations Shapley cited as evidence that the spirals were nearby didn't exist. Van Maanen was simply wrong. The size of the Milky Way, at 100,000 light-years in diameter, is more like Shapley's estimate than Curtis's—but clearly, that didn't mean the nebulae had to fall within it. If Curtis had known for certain at the time that the Milky Way was huge, he might have doubted all the other lines of evidence—incorrectly. "This debate illustrates forcefully," Shu continues, "how tricky it is to pick one's way through the treacherous ground that characterizes research at the frontiers of science." Michael Turner, a cosmologist at the University of Chicago, said essentially the same thing in the early 1990s, just before the COBE satellite made a first pass at measuring fluctuations of the CMB: "If you design your theory to fit all the data we have right now . . . then the theory is almost certainly wrong, because it's very unlikely that everything we think we know about the universe at this moment is correct."

But as instruments become more powerful, we know more correctly what the universe actually looks like. Not only do spurious measurements go away, and arguments about such things as the size of the Milky Way get resolved, but strange observations that don't seem to fit any known theory start falling into place. One of these, noted by both Curtis and Shapley in the debate, was the fact that virtually all of the spiral nebulae seem to be rushing away from Earth. The source of this finding was Vesto Slipher, an astronomer at Lowell Observatory in Flagstaff, Arizona. The observatory had been built by Percival Lowell, a rich Bostonian who wanted to prove the hypothesis that there were canals on Mars, but Slipher was more interested in using a spectrograph to understand the composition of heavenly objects.

The spectrograph could also tell him something else: if an object is moving toward the Earth, its light, and therefore its spectrum, and therefore all the dark lines that interrupt that spectrum, will show up in the telescope with a higher frequency than they started out with. They'll be bluer. If the object is receding, everything gets a bit redder. That's due to the so-called Doppler shift: approaching light waves pile up on one another, shortening the distance between their peaks, while receding waves stretch out (sound waves do the same: listen for an abrupt drop in pitch as an ambulance with a wailing siren races by).

If you measure the shift in the lines, you can calculate the speed of approach or recession in a star or a nebula, and by 1914, Slipher had found that while Andromeda is coming at us at about 300 kilometers per second, the rest of the score or so of nebulae he looked at are flying away, some of them at more than three times that speed. Though he didn't realize it, Slipher had discovered the expanding universe. So while Shapley and Curtis based their arguments on faulty observations and theories, they also had in hand some observations that were not only correct but profoundly important in describing the nature of the universe. They just had no way of knowing it.

Within a decade, however, they would—thanks to Edwin Hubble. By the 1930s, Hubble would be as much a celebrity as a scientist, hobnobbing with fellow southern Californians such as Charlie Chaplin, Helen Hayes, William Randolph Hearst, and Aldous Huxley. But in the early 1920s he was known only within the astronomical community as the lucky man who had his hands on the world's most powerful telescope: the 100-inch Hooker on the summit of Mount Wilson. The Hooker, which took twelve years to build, had lately replaced the 60-inch reflector Shapley had used for his own research. Now Shapley was moving on to Harvard, and Hubble was brought in from the Yerkes Observatory in Wisconsin to take over the new instrument.

Hubble's journey to Mount Wilson was roundabout. He'd studied science at the University of Chicago while playing basketball and boxing on the side (evidently he was so good at the latter that several fight promoters tried to get him to turn pro). But when he went to Oxford on a Rhodes scholarship, he studied law instead, to please his father. When Hubble returned in 1913, he took a job as a high-school Spanish teacher in New Albany, Indiana—but a year later, he returned to his first love. Hubble became a graduate student at Yerkes and began to study the spiral nebulae. His astronomical career was interrupted one more time, by World War I. Hubble spent most of the war at Aberdeen Proving Grounds in Maryland, using his knowledge of physics to help design and test artillery.

But finally, in 1919, he arrived at the mountain. "In his mind's eye," writes Gale Christianson in his exhaustive 1995 biography, "Hubble was still a major, an image he cultivated from his first day on the mountain. His usual attire was a shirt, tie, Norfolk jacket, jodhpurs, and high-topped military boots." He'd also kept the "Oxford mannerisms" of speech, as Christianson calls them, that had charmed his female students in New Albany but struck some other astronomers as excessively pretentious. He was, nevertheless, a serious observer. Spending all night in frequently subfreezing temperatures, Hubble turned the Hooker on the spiral nebulae, classifying them into different types based on their structure, and looking for clues as to their true nature.

In 1923, he found one. One of Curtis's arguments for the proposition that the nebulae are extragalactic was the presence of what appeared to be novae within them. In that year Hubble spotted three in the haze of the great nebula in Andromeda, more formally known as M31. When Hubble looked at older photographs of M31, however, he found that one of the three showed up in some of these as well. This was clearly not a nova, which even then was understood to be some sort of exploding star. It was instead a so-called Cepheid variable star, and its discovery meant the Great Debate was essentially over.

Cepheid variables were at the time, and remain today, among the most valuable tools in astronomy. That's because the sky looks to modern observers, as it did to Galileo and Copernicus and the ancient Greeks and even the Neanderthals, as though it were two dimensional. There aren't any obvious depth cues (except on the rare occasion when the moon or the Sun passes in front of a star or planet). A star that looks very bright, like Sirius, could either be intrinsically brilliant and far away, or just moderately bright and close by (Sirius happens to be the latter). The same goes for any other heavenly object. That's the reason nobody could say for sure whether the spiral nebulae were inside the Milky Way or far beyond it.

If all stars had the same inherent brightness, though, it would be easy to determine their distance: the bright-looking ones must necessarily be closer. That's the case with Cepheids. In 1908, Henrietta Leavitt, a research assistant at Harvard College Observatory, made a study of Cepheids and realized that there was a direct relationship between the regular pulsation in a Cepheid's brightness and its absolute luminosity. The longer the period, the brighter a Cepheid inherently was. Find two Cepheids with the same period, and the dimmer-looking one must be farther away by a distance that could be precisely calculated. The study was ignored, in part because the researcher was a woman and thus unqualified to be a "real" scientist. And while her more extensive 1912 report on the Cepheid's period-luminosity relationship was finally accepted and used by observers to gauge cosmic distances, Leavitt was ordered by her boss at Harvard to move on to a different project. She had made a powerful and lasting contribution to astronomy (Cepheids resolved by the Hubble Space Telescope were used a few years ago to measure accurately the distance to the Virgo Cluster of galaxies for the first time), but she didn't have enough status to be allowed to follow her own research interests.

Hubble's new Cepheid had a period of 31.4 days, which translated into a distance of about a million light-years (the true distance is more like 2 million; early Cepheid measurements were relatively crude). And whether you accepted Curtis's 30,000-light-year-diameter Milky Way or Shapley's 250,000, this was far beyond its edge. Hubble wrote to Shapley reporting the discovery. Shapley's immediate response, according to Katherine Haramundanis in her biography of Cecilia Payne Gaposchkin, an early female astronomer (not merely an assistant), who was present when he opened it: "Here is the letter that has destroyed my universe." Shapley actually held out hope there might be some sort of error. But within six months, Hubble had found more than a dozen more Cepheids, some in Andromeda and others in the nebulae M33, M101, and NGC 6822. All of them were well outside the Milky Way; their host nebulae were unquestionably galaxies in their own right.

That discovery transformed humanity's understanding of its place in the cosmos. Not only was the Earth not the center of Creation, as Copernicus had realized in the late 1500s, but now the Milky Way was only a small part of a much greater universe. But Hubble's next discovery would be even more surprising, and so baffling that it would take a third of a century to be certain of its astonishing implication. Since the midteens, Vesto Slipher's measurements of nebular redshifts had been a nagging mystery. The nebulae were essentially all speeding away from us; the fact that they were distant galaxies made this mad rush even stranger. Slipher and other astronomers speculated that there might be some relationship between a galaxy's recessional speed and its distance, but there was no hard evidence for this idea. Because Slipher's telescope at Lowell Observatory was relatively small, he couldn't pursue it further.

Hubble, with the 100-inch Hooker and with his increasing skill at measuring the distances to galaxies, could. He hired Milton Humason, a muledriver with an eighth-grade education turned observatory janitor turned volunteer observing assistant. (Humason was so skilled at taking photographs through the telescope that his were the first images of Pluto; unfortunately, the planet was too blurry to notice in his 1919 pictures, and the formal discovery would not come for another eleven years.) Under Hubble's direction, Humason began taking unprecedentedly fine images of the nebulae, looking for Cepheids to gauge their distance. By 1927, they had measured the velocities and distances of about fifty galaxies. When they plotted one against the other, the dots were strung out on an impressively straight line. Slipher's speculation was correct: the farther away a galaxy was, the faster it was moving. This relationship, which would eventually be known as Hubble's law, suggested one of two things: either the Milky Way was at the center of the universe, with every other galaxy speeding away from it—a preposterous idea—or the entire universe was expanding uniformly, growing at a constant rate in all directions at once; still very hard to swallow, but at least in keeping with the Copernican principle that says we don't occupy a unique position. (This principle had lately been reinforced by the proof that, contrary to what Curtis had believed, the Sun is nowhere near the center of the Milky Way.)

It may not be intuitively obvious why a distance-velocity relationship suggests overall expansion, but it's simple enough to explain. Take the analogy of a large rubber band with a series of marks one inch apart. One mark represents the Milky Way; the others are other galaxies. Now stretch the rubber band to twice its original length in, say, one second. The mark nearest to the Milky Way is now twice as far away; it's traveled an inch, with a speed of one inch per second. The second mark is also twice as far from the Milky Way; it's gone from two to four inches in that same second, so its speed is two inches per second. The fourth mark has gone from four to eight inches—four inches per second. And so on. From the point of view of the Milky Way, the other galaxies are moving away, in both directions, and the farther away a given galaxy is, the faster it's receding. But since the whole rubber band is stretching, the point of view of a different dot will be exactly the same: everything else is moving away from it, with the same distance-velocity relation.

What an expanding universe actually signified was unclear. Hubble himself refused to speculate on possible implications; he was purely an observer, not a theorist. Theorists, however, had been considering the possibility of an expanding universe for more than a decade. Albert Einstein was the first: after creating general relativity, which recast gravity not as a force but as a warping in the geometry of space and time, he tried applying its equations to the entire universe. He was surprised to find that space should be expanding or contracting, but that it had no option to remain static. Astronomers, who hadn't yet incorporated Slipher's observations into their collective wisdom, claimed that it was static nevertheless. Distressed that his beautiful equations didn't describe the real universe, Einstein grudgingly added an extra factor he called the "cosmological term." It corresponded to a physical force that kept the universe propped up against collapse—an outward pressure that would precisely counterbalance the effects of gravity. On esthetic grounds alone, Einstein hated it. Then came Hubble, and suddenly Einstein was free of the cosmological term. Calling it the greatest blunder of his life, he excised the offending force from his equations and went to visit Hubble on Mount Wilson, peering through the telescopes and terrifying the local staff by clambering up the 100-inch telescope's steel framework.

Einstein had, uncharacteristically, allowed observations to overrule his elegant theory. (After general relativity passed its first observational test, during an eclipse of the sun in 1919, he was asked how he would have felt had it failed. His answer: "I would have felt sorry for the dear Lord.") While he stopped working on cosmological models that suggested an expanding universe, however, others did not. Without the cosmological constant to constrain them, the equations of general relativity can produce all sorts of universes, depending on what assumptions you make. The Dutch physicist Willem de Sitter explored a universe that expanded but had no matter in it. Aleksandr Friedmann, a young Russian physicist working at the University of Petrograd, argued that Einstein's cosmological term, also known as the cosmological constant, didn't produce a static universe in any case, and went on to describe a cosmos that expanded from an initial point of zero radius.

But it was Georges Lemaître, a chubby and, by all accounts, prickly Belgian priest who took the idea farthest. Lemaitre, who had studied mathematics and physics at the University of Louvain before taking his vows, first published his theory in an obscure Belgian journal in 1927: it argued for an expanding universe and predicted that the redshifts of the nebulae would increase with distance. Nobody noticed, but in 1931 Sir Arthur Eddington, the British astronomer who had confirmed general relativity in 1919, took note of the idea and had it republished in English. Lemaître went on to expand his idea, suggesting that the expansion began when a gigantic "primeval atom" exploded, flinging outward the matter that would eventually form the modern universe. "The evolution of the world," he would write in his 1950 book, *The Primeval Atom: An Essay on Cosmogeny*, "can be compared to a display of fireworks that has just ended: some few red wisps, ashes and smoke. Standing on a well-chilled cinder, we see the slow fading of the suns, and we try to recall the vanished brilliance of the origin of worlds." Lemaître's was the first statement of a cosmological model that would eventually be known as the Big Bang. (The term was formally coined in 1950 by British astronomer Fred Hoyle, who didn't buy the theory. It's generally believed that "Big Bang" was a pejorative label for what he thought was a silly idea, though Hoyle would claim some forty-five years later that he hadn't meant it that way at all. It is interesting to note that Eddington, who didn't approve of a beginning to the universe either, had written back in 1928, before Lemaître even suggested the primeval atom: "I simply do not believe that the present order of things started off with a bang . . . the notion of an abrupt beginning to this present order of Nature is repugnant to me.")

Conventional wisdom has it that the next significant event in understanding the early universe was the theoretical prediction that a Big Bang implied a glow of leftover microwaves—but, in fact, this cosmic microwave background was detected nearly a decade before that prediction. And just as with Vesto Slipher's detection of the expanding universe, nobody had any idea of its significance—nor would they for another quarter of a century. In the late 1930s, an astronomer named S. W. Adams, working at Mount Wilson, was making careful measurements of the spectra, not of galaxies, but of individual stars. He found that a particular wavelength of starlight in the star Rho Ophiuchi was being absorbed by molecules of cyanogen gas drifting in interstellar space. Some of these molecules were spinning, a phenomenon that happens when the gas is heated, and by comparing the number of spinning to nonspinning molecules, Adams and a Canadian astronomer named Andrew Keller were able to calculate its temperature. It was about 2.3° C above absolute zero, they figured. One possible explanation was that interstellar space was being bathed with electromagnetic radiation of about that temperature—but there was no plausible source for such radiation, they thought. Another idea, less convincing but more believable, was that electrons, stripped from their atoms in the atmospheres of stars, were zipping through space and occasionally slammed into cyanogen molecules, sending them whirling like pinwheels. It was a mystery, but, as far as anyone knew, a minor one, and it was soon forgotten.

Meanwhile, unmindful that the CMB had been detected, albeit indirectly, theorists were moving slowly toward predicting its existence. The first step in that direction came from George Gamow, a Russian physicist who emigrated to the United States in 1933. Gamow had studied under Aleksandr Friedmann in Petrograd, and also with Ernest Rutherford, who discovered the structure of the atom, and Niels Bohr, who helped found modern quantum theory. (Absurdly, he would later be denied security clearance to work on the Manhattan Project because he'd been commissioned as a major in the Soviet Army—but only so he could teach at a military college, and despite the fact that the USSR had subsequently sentenced him to death in absentia for defecting.) Shortly after Gamow arrived in this country, he began pondering the relative abundances of chemical elements in the universe. By using their spectroscopes, astronomers had determined that stars are about 75 percent hydrogen, the lightest element known, and 23 percent helium, containing only a tiny amount of the remaining several score chemical elements (none, however, heavier than iron; gold, silver, lead, and other heavy metals were found only on asteroids, meteors, and the planets). Atomic physicists had discovered, meanwhile, that in a sense the elements are all elaborations on hydrogen. A hydrogen atom is simply an electron orbiting a proton; every other element is made up of multiple electrons orbiting a core of protons and neutrons.

Gamow wondered how the heavier elements might have been manufactured from these building blocks, and where and when this might have happened. The best place, he guessed, was the very early universe. If the cosmos was expanding, as Hubble had made it pretty clear it was, then it must have been smaller and denser in the past. Go back far enough, to a point where matter was squeezed into a relatively small volume, and the temperature must have been very high—a phenomenon so basic it's essentially a law of physics. In such a superheated environment, he figured that protons, neutrons, and electrons would be flying around and, at the right temperatures, sticking together to form hydrogen, then helium, and on up the periodic table of the elements. He called the hot, preelemental ball of gas "Ylem," from the Greek word for the primordial substance of the cosmos.

During the 1940s, Gamow published several papers on this idea. One of them demonstrates his addiction to practical jokes: Gamow wrote a paper with his student, Ralph Alpher, elaborating the formation of the elements. On submission to the journal *Physical Review*, though, he added the physicist Hans Bethe as a third author. Bethe had nothing to do with the research or the writing, but having the names "Alpher, Bethe, Gamow" at the top made the byline into a pun; the first three letters of the Greek alphabet are alpha, beta, and gamma. Bethe had pulled a few hoaxes himself during his career, and when the journal's editors contacted him with the news that his name was listed as an author "in absentia," he agreed to go along with the scheme.

Although his motivation wasn't to understand the birth of the universe, Gamow was the first to try and model its detailed physics. He wasn't the one, however, who predicted the existence of the cosmic microwave background. That fundamental insight would come from Alpher and Robert Herman, who, like Gamow, worked at the Johns Hopkins Applied Physics Laboratory in Baltimore in the years after World War II. First, in a series of papers, the pair worked out in detail Gamow's ideas about the creation of the elements; as it turned out, his scheme worked only to manufacture the first few, lightest elements of the periodic table before petering out. But then they went on to analyze the characteristics of the Ylem and made a prediction that would ultimately lead to the Big Bang's nearly universal acceptance. Because of its extraordinarily high density, the Ylem's fierce heat would have been distributed evenly at any given time. In the language of thermodynamics, it was a "blackbody," so called because the idealized theoretical model for a body in thermal equilibrium is an object so black that it absorbs all energy that falls on it, reflecting none.

Physicists had known since the late 1800s that a blackbody, whether it's in solid, liquid, or gaseous form, emits electromagnetic radiation. The wavelength, or color, of that radiation depends only on temperature, with the coolest objects sending out the longest wavelengths. The Sun is yellow-white because its surface temperature is about 6000 Kelvin (that is, 6000° Celsius above absolute zero, which, at −273.15° C or −459° F, is the coldest possible temperature). A red star like Betelgeuse is red because its temperature is more like 3000 K. If you heat a vat of molten metal to 3000 K, it will have precisely the same color as Betelgeuse; send it to 6000, and it will look like the Sun. Now cool the metal to room temperature. It still emits radiation, but now the "light" is infrared, with too long a wavelength for the eye to see. The material doesn't matter, just the temperature—a bowling ball at room temperature, or a wheel of Jarlsberg cheese, or a statuette of Elvis, all will emit the same wavelength of infrared radiation as the cooled metal and as each other. (To be more accurate, they'll emit the same mix of wavelengths. The color we see or detect is that of the most abundant wavelength, but blackbody radiation at any given temperature includes other, less prominent wavelengths as well. Plot these wavelengths on a chart and they'll describe a curve: the Sun's curve peaks at a wavelength of about 5000 angstroms and falls off rapidly on either side of that.)

Alpher and Herman understood that as the Ylem expanded and cooled, its temperature, and thus its color, would constantly be dropping, from the supershort wavelengths of gamma rays down to X-rays down to ultraviolet light. At temperatures high enough to generate these energetic forms of light, atoms can't form fully; any electrons that tried to combine with an atomic nucleus would instantly be stripped away again. Photons, the particles that carry electromagnetic energy, would be able to travel only a tiny fraction of an inch before they would run into one of these free electrons, be absorbed, and then be reemitted in a random direction. The light of the Ylem would thus have been something like an incandescent fog—until the temperature dropped to about 6000 K. At that point, electrons would have been able to combine with nuclei at last; with the electrons taken out of circulation, the light generated at that temperature would shine freely.

With nothing to block it, in fact, Alpher and Herman realized that this light should still be shining. But as the universe expanded, the light traveling through it would have expanded as well; what started out as visible light should by now have been stretched like taffy, its wavelength pulled out through the red end of the visible spectrum, into the infrared and beyond. Today, they calculated, it should be somewhere in the microwave region, the wavelength characteristic of a blackbody hovering at about 5°C above absolute zero—the still-lingering flash of the primeval atom, streaming in from all directions in the sky.

This prediction, published in the prestigious British journal *Nature*, offered a powerful test for Lemaître's cosmology. The expanding universe was certainly consistent with the priest's assertion that the universe had an explosive start. But that was hardly surprising, since the theory used the expanding universe as its fundamental assumption. To be truly convincing, Lemaître's model had not only to explain the known facts, but also to predict some still-unknown fact about the universe that scientists could go out and measure. That's what it had taken for Einstein's general relativity to be accepted: the theory explained planetary motions elegantly—better, even, than Newton's theory of gravity could do. But it was only when observers saw that starlight was warped as it passed by the Sun, as Einstein predicted, that physicists accepted relativity as more than a clever mathematical construct.

Now Alpher and Herman had made an equally bold theoretical prediction, which could help settle an equally profound question about the universe. The response of observers was mostly to ignore it. In retrospect, this seems absurd—who wouldn't want to get credit for solving perhaps the greatest outstanding cosmic mystery? At the time, though, the decision to try and measure the cosmic background radiation wasn't exactly a no-brainer. As they recall in their 2001 memoir, *Genesis of the Big Bang*, Alpher and Herman talked to radio astronomers about looking for the cosmic microwave background radiation. They also presented their results at meetings of the American Physical Society twice, and in one case the presentation led to a press release that was picked up by newspapers across the country.

Nobody elected to test the prediction. The consensus was that it wouldn't be possible to detect; the signal would be too faint to pick out of the microwave noise from the atmosphere and cosmic rays and stars. Gamow himself talked a graduate student named Joseph Weber out of attempting the detection as a thesis topic, on the grounds that it was impossible. He wasn't even convinced that the CMB radiation must really exist, despite Alpher and Herman's calculations. (He would come around eventually, however.) Alpher and Herman later speculated that if they'd plotted the blackbody curve of the microwave background radiation in their paper rather than just describing it, the graphical presentation would have inspired more interest. Another reason the prediction was mostly ignored was the failure to make significant amounts of anything heavier than helium in Gamow's hot early universe. With the original premise of the theory invalidated, the rest was taken less seriously. (It would be Fred Hoyle, the Big Bang's most ardent opponent, who would later be among those who showed that most elements are created inside stars.)

In the end, both Gamow and the general consensus were obviously wrong. What's less obvious is that Alpher and Herman might have had their prediction tested right away if they'd talked to the right people. In fact, as noted above, it had already been tested by McKeller and Adams's work on interstellar cyanogen—but Alpher, Herman, and Gamow were physicists; they didn't follow the astronomical literature. They also didn't, and couldn't, follow every subspecialty of physics. They didn't know that a colleague named Robert Dicke had already gone out and tried measuring the CMB, albeit inadvertently. Dicke was unquestionably among the most brilliant physicists of the twentieth century, equally creative as a theorist and an experimentalist, though the general public probably wouldn't recognize his name. Among other things, he helped invent radar during World War II, did much of the groundwork that led to the laser, and promulgated a theory of gravity that was for a decade or more considered a serious contender to topple Einstein's general relativity. Dicke thought of having the Apollo astronauts put three-corner reflectors on the lunar surface so Earthbound scientists could bounce lasers off them and measure the Earth-Moon distance with a precision of centimeters. (As recently as the spring of 2002, these measurements led geologists to conclude that the Moon has, against all expectations, a liquid core.) Dicke even gets credit for stumbling on the ballpoint pen, invented in South America, and having it imported to the United States for the first time.

In 1946, while working to perfect radar, Dicke invented and built a device called a radiometer, designed to measure the returning pulses when a radar beam bounced back from an enemy plane or ship. Worried that water vapor in the atmosphere might dilute the signal, he set out to test the radiometers. "While we were at it," recalled the now-deceased Dicke in a 1991 interview, "we pointed them at the sky, both in Massachusetts and Florida." He figured that he'd be able to see celestial microwaves—not from the Big Bang, since he wasn't even thinking about cosmology in those days, but from galaxies. The radiometers were sensitive enough to see microwaves as cool as 20 Kelvin, and saw nothing. But it wouldn't have been all that difficult to increase the sensitivity if Dicke had seen any reason for doing so. Absent a conversation with Alpher and Herman, he didn't.

Another reason for the experimental community's lack of response was that the still-unnamed Big Bang had at least one significant problem and one apparently fatal flaw. The problem was that it required the universe to have a beginning, a concept as difficult to swallow for most people as it had been for Sir Arthur

Eddington two decades earlier. The pope, by contrast, loved this idea, since it fit so nicely with Genesis 1:1. In 1951, Pius XII put his stamp of approval on the theory. The same Church that put Galileo under house arrest for asserting that the planets orbit the Sun was now fifteen years ahead of the scientific community in accepting the Big Bang. The fatal flaw was that the Big Bang suggested that the universe was younger than the Earth. Edwin Hubble had plotted the speed of galactic recession against galactic distance and come up with an overall expansion rate, which had come to be known as Hubble's constant. Using that constant, it was nothing more than a high-school algebra exercise to calculate backward and see when all of the galaxies (or the matter they were made from, anyway) had begun their outward journey. The answer was 2 billion years. Unfortunately, geologists had determined from the decay of radioactive minerals that the Earth was more than 4 billion years old. And geology, unlike cosmology, was a firmly established branch of science.

It would become clear a decade or so later that while Hubble's measurements of the galaxies' recessional velocities were accurate, his distance measurements were way off—he thought the galaxies were much closer than they actually were, which in turn had skewed his constant badly. But during the 1950s, the Big Bang acquired both its popular name and its chief rival from the same source. Fred Hoyle was a brilliant young physicist, easily as creative and iconoclastic as George Gamow, who would go on to help explain in detail how many of the chemical elements are forged in the cores of stars rather than in the Ylem, as Gamow had proposed. He would also become a successful author of science fiction. And, like Gamow, he would become the coauthor and chief representative of a leading version of cosmology. According to one story, Hoyle and two University of Cambridge colleagues, Herman Bondi and Thomas Gold, went to see a movie called *Dead of Night*, in which the story ends at the point where it begins. Inspired by this bizarre plot, Hoyle conceived (again, according to the legend), and the three worked out, the so-called Steady State theory of the universe.

The Steady State theory acknowledged the inescapable fact that the galaxies were moving apart. But in contrast to the Big Bang, it argued that they'd always been doing so. This was in keeping with the so-called Copernican principle. A half-millennium ago, Copernicus argued that the Earth is not the center of the universe, and in the centuries since, astronomers had gradually realized that the Sun isn't the center, and, finally, with the discovery of the expanding universe, that our galaxy isn't at the core of things, either. The principle states that we occupy no special vantage point in space; according to the Steady State theory, we also occupy no special place in time. The universe has always looked pretty much as it does now. In order to explain the fact that the galaxies aren't all infinitely far apart by now, the Steady State further argues that matter is continuously being created in empty space, and that this matter eventually gives rise to new galaxies, which fly apart in turn. The Steady State wasn't any more preposterous than the Big Bang, really: the latter suggested that all the matter in the universe was created in an instant; the latter only called for the creation of one hydrogen atom per cubic meter of space per year, on average.

CHAPTER 3

A Whisper of Microwaves

Through the 1950s, both the Steady State and the Big Bang had plenty of proponents, although, as noted in the previous chapter, the Big Bang alone offered a definitive test. It had already passed that test with the observations of interstellar cyanogen, and could have passed again if someone had pointed a Dicke radiometer at the sky to look for it. And although nobody realized it, the prediction of cosmic microwave background radiation would come within a hair's breadth of confirmation several more times during the 1950s and early 1960s. Alpher and Herman, who were naturally frustrated by the failure of observers to take their prediction seriously, would find out decades later that a Russian radio astronomer named Tigran Shmaonov had detected the CMB with a captured German radar antenna. A French radio astronomer named Emile Le Roux saw it and estimated the temperature to be about 3° Kelvin—within 10 percent of the actual value. An American astronomer named William Rose measured it and came up with a value of between 2.5 and 3 K. It's doubtful that any of these scientists were aware of Alpher, Herman, and Gamow's work, since observational astronomers don't tend to spend much time studying up on theoretical physics. In any case, they didn't realize what they were seeing, and didn't publish their work in any journal a theoretical physicist would be likely to read.

The closest near-miss of all happened in the early 1960s, at Bell Labs, in Holmdel, New Jersey. Before AT&T was broken into pieces by court order in the 1980s, the company monopolized virtually all telephone service in the United States. It had so much money that it could afford to run a laboratory devoted to pure research, on the principle that letting very smart scientists do whatever intrigued them could lead to important, though not always telephone related, work. The transistor, which made modern computing possible, was invented at Bell Labs. So was the science of radio astronomy, and the laser.

One of AT&T's ventures involved trying to bounce microwave telephone signals off a high-altitude balloon known as Echo. The receiver was a huge, horn-shaped antenna, twenty feet across at its open end. Because the reflected signals were so weak, Bell Labs engineer Edward Ohm was assigned in 1961 to characterize all of the excess noise that might contaminate an actual microwave signal. He accounted for all of it, but he almost certainly misassigned a few degrees' worth of that noise to natural microwaves from the warm ground. In 1964, a couple of Russian physicists, A. G. Doroshkevich and I. D. Novikov, looked at Ohm's paper in a deliberate attempt to confirm Alpher and Herman's prediction. They bought his explanation: no excess noise unaccounted for, so no evidence of the microwave background. If they'd been radio astronomers instead, they might have realized that a cosmic signal could easily have been confused with emanations from the ground, and come to a dif-

The horn antenna used by Arno Penzias and Robert Wilson to find the cosmic microwave background (CMB) accidentally in 1964 was originally built to detect signals bounced from the Echo satellite.

ferent conclusion. Another Bell Labs engineer, W. C. Jakes, re-analyzed the horn antenna for noise in 1963 and decided there actually was about 2.5° worth of noise that couldn't be explained. But that's as far as he went. If the Russians had seen that paper, they would probably have discovered the CMB.

Finally, in 1964, two Bell Labs radio astronomers, Arno Penzias and Robert Wilson, decided they were going to get to the bottom of the noise problem, once and for all. They would find out where the excess was coming from. To get the most accurate measurements possible, they built a device called a "cold load," a reference source of microwaves at a very cold temperature that gives you something to compare with the signal coming from the sky. Penzias and Wilson were fanatical about getting rid of or accounting for every identifiable source of excess noise; they even trapped some pigeons who had nested in the antenna, and cleaned out the "white dielectric" material the birds had left behind. About 3° of noise was still there. It was evidently coming from some source in the sky—a source not localized to our galaxy, but appearing to come from all directions at once. Like other radio astronomers before them, Penzias and Wilson had no idea that Alpher, Herman, and Gamow had predicted such radiation, so they simply noted it.

Penzias and Wilson also had no idea that only thirty miles away, two young faculty members at Princeton University, Peter Roll and David Wilkinson, were busy constructing a microwave antenna and a cold load on the roof of Guyot Hall, the geology building next door to Palmer Physics Lab, on the Princeton campus. The two experimentalists were acting on the suggestion of their mentor, Bob Dicke, who had, some two decades after his radar experiments, decided to look for the cosmic fireball after all.

Dicke wasn't trying to confirm Alpher and Herman's prediction. He'd never heard of it, or so he would later recall. Dicke had come up with the idea of a fireball more or less independently, by an entirely different route from the one Gamow had taken. He had been exploring the idea of a cyclical universe, in which phases of expansion and contraction alternated, like the pumping of an accordion. But because stars were continuously converting hydrogen and helium into heavier elements, he had to explain why most of the universe, after so many cycles, was still made of these lightest of elements. The answer, he decided, was that the universe got so hot during its contraction phase that the heavy elements were deconstructed into light ones again. And if it got that hot—about 10 billion degrees, he figured—the signature of the last compression cycle should be a cosmic

The apparatus built by David Wilkinson and Peter Roll atop
Guyot Hall in Princeton was ordered up by their colleague Bob
Dicke to look for the CMB deliberately, but it wasn't finished
in time; the Princeton scientists used it later that year to confirm the Bell Labs discovery. Wilkinson is fully visible; Roll is
obscured by the equipment.

microwave background. He and his colleagues—Wilkinson,
Roll, and a theorist named Jim Peebles who calculated the likely
temperature—were "deficient in not reading the literature very
carefully," Dicke would admit in a 1991 interview. He also acknowledged that he'd attended a talk by Gamow at Princeton
on the cosmic fireball. But either Gamow didn't mention the implication of a cosmic microwave background, or Dicke—scrupu-
lously honest, according to everyone who knew him—had regis-
tered the idea only in his subconscious.

Wilkinson had arrived at Princeton two years earlier, fresh
from graduate school at the University of Michigan. He didn't

know much about cosmology, or even astronomy at that point. In fact, he had started out planning to be an engineer, not a physicist. In his first year as an undergraduate at Michigan he'd been aiming toward engineering. His father hadn't even gone to college, but, said Wilkinson in a conversation shortly before he himself died in early September 2002, "he could pretty much fix anything. He's the reason I became an experimentalist, a hands-on scientist." Fortunately for the world of physics, one of Wilkinson's first engineering courses was titled "Cement." One entire semester's worth. "That course made me realize how much I enjoyed physics," he said, so he switched his ambitions slightly and ended up with a degree in nuclear engineering. He went on to graduate school where he did a Ph.D. in particle physics. "My thesis project," he said, "involved measuring how strong a magnet the electron is." Several well-established research groups were trying to do the same thing, but Wilkinson's measurement was the most accurate. "That let me write my own ticket," he said. The destination he chose was Princeton, where Bob Dicke was working on at least six important projects at once.

For the first couple of years, Wilkinson built experiments to test Dicke's theory of gravity. Then, in 1964, Dicke got Wilkinson, Peebles, and Roll together and suggested they try and find this cosmic radiation he'd been thinking about. "He drew a picture of the apparatus on the board," said Wilkinson, "and said, 'OK, boys, go build it.' " Wilkinson, at least, didn't expect at first that they'd find anything. "It just seemed too remote and fantastic that there would be radiation left over from 14 billion years ago, and even more fantastic that if it were, nobody had seen it."

In the spring of 1965, a couple of months before Wilkinson and Roll expected to get their experiment up and running, Jim

Peebles gave a talk at a conference of the American Physical Society in New York, in which he described the Princeton detector and its theoretical rationale. In the audience was a radio astronomer named Ken Turner—like Peebles, a former graduate student of Bob Dicke's. Turner described the talk to a colleague, Bernard Burke, who was then at the Carnegie Institution of Washington. About a month later, Burke got a call from another colleague: Arno Penzias. They talked about several things, but eventually the conversation came around to the excess noise in the Echo horn antenna. Burke immediately thought of Peebles's talk and urged Penzias to give the Princeton group a call.

Wilkinson, Peebles, and Roll were having their lunch as usual in Dicke's office when Penzias's call came in. They didn't pay much attention, Wilkinson recalled—Dicke was always getting calls—until they heard the words "cold load" and "horn antenna." Dicke hung up, looked at them, and said, "Well, boys, we've been scooped." The Princeton group piled into a car, drove over to Bell Labs, and had a long conversation with Wilson and Penzias. The upshot would be back-to-back papers in the *Astrophysical Journal*, one from Bell Labs announcing the discovery, the other from Princeton explaining it. Nobody realized at the time that Wilson and Penzias would win the 1978 Nobel Prize for the detection, but Wilkinson doesn't seem to hold a grudge at losing out on the biggest honor in science. "They built a damn good cold load," he said, "and they had an excellent receiver. Wilson is an ace. One of the best instrument builders I know. A lot of people would have given up, said 'the hell with it—we've looked for bird crap and we've done the best we can.' They didn't give up."

He didn't remember being annoyed at the time, either. "A little disappointed, maybe, but there was a lot of work to do." In fact, the intital detection didn't really prove the existence of the CMB at all. If Penzias and Wilson had seen true blackbody radiation, it would not only have a peak in the right place—which is what they did see—but also a lower intensity of emission near the peak. The Bell Labs astronomers had found the highest point on what should be a curve. Now the job was to find other points, and they'd better fall on the curve as well—and that's what Wilkinson and Roll went out and did. The initial discovery made the front page of the *New York Times*. "Signals Imply a 'Big Bang' Universe," said the headline. (Dicke's original idea about a cyclic universe had to be abandoned because the cycles would get shorter as you went into the past; ultimately, you'd still have had to start with a beginning of time, and thus a Big Bang.)

Despite the publicity, however, Wilkinson remembers that it took a while for other scientists to accept the detections. "Physicists didn't really understand cosmology back then, so they couldn't really evaluate the science." In fact, the initial paper in which Dicke's group explained what Penzias and Wilson had seen was almost laughed at. "It was at least a year," he said, "before people stopped giving us a hard time when we went out and gave talks. We were convinced of what we'd seen. But the measurements are so intricate you have to really think about where you might go wrong."

Over the next few years, though, Wilkinson and others would substantially erase those doubts by measuring the CMB at several wavelengths. Astronomers Neville Woolf and George Field, meanwhile, would finally make the connection that the long-ago detection of excited molecules of interstellar cyanogen was independent evidence of a bath of cosmic microwaves. By the late 1960s, the Steady State theory of cosmology had been abandoned by pretty much everyone except Fred Hoyle and a couple of colleagues. The Big Bang became cosmology's Standard Model, the basic scenario into which subsequent observations would be fit. It was a very simple model at first: all it said was that the universe began in a state of extreme heat and density, then evolved under the influence of gravity.

It was clear almost from the beginning, however, that this simple story was too simple. It couldn't explain the present-day cosmos in any realistic detail. One problem was that in order to end up with galaxies and clusters of galaxies and lots of empty space in between, you had to start with some regions of excess density right from the beginning. A Big Bang that was too perfectly smooth and uniform couldn't give rise to the universe we see. Without some "seed," some region of slightly higher gravity than average to start drawing in matter, the process would never begin. No one had a clue, though, whether these regions actually existed in the early universe, and if so, why. One early suggestion was that the knots of high density were the result of turbulence, but experts in turbulence said that wasn't very plausible.

Another major unanswered question concerned the geometry of the universe. Einstein's general theory of relativity had reimagined the universe as a dynamic physical object rather than an empty stage. One attribute of this cosmic object was curvature: the universe must be positively curved, like the surface of a sphere, or negatively curved, like the surface of a saddle, or flat, like a sheet of paper. These oft-cited analogies aren't perfect, however. The surface of a sphere or a saddle or a sheet is two dimensional, but it exists in a three-dimensional world. You'd think that their three-dimensional analogues would have to exist in a four-dimensional world. But they don't. The Riemannian geometry Einstein used to construct relativity suggests that the universe should be curved without requiring it to be embedded in a higher-order space.

Einstein also argued that mass and energy warp spacetime. So the thing that governs which sort of curvature the universe will have is the density of mass (or its equivalent energy) it contains. If the density is high, the universe will have positive curvature. If it's low, the curvature will be negative. If it's on the borderline, the universe is flat. As a shorthand, cosmologists use the Greek letter omega to stand for this mass/energy density. An omega greater than one means high density; less than one means low, and an omega of precisely one means a flat universe.

The trouble was that astronomers had been measuring the density of matter in the universe, and it suggested that omega was about 0.01, one percent of the amount that would make the cosmic curvature precisely flat. (With Einstein's cosmological constant long since abandoned, there didn't seem to be a significant energy component to omega at all.) But that seemed very odd, as Bob Dicke first pointed out in a talk in 1969. While a factor of 100 seems pretty large, omega could have turned out to be anything. It could have been 0.00000001. It could have been 1,000,000,000. The fact that it's so comparatively close to 1 seemed like too much of a coincidence. And it gets worse. The nature of omega is such that it tends to move away from 1 as the universe expands. To be 0.01 today, it would have had to be 0.99999999999999999999999999999999999999999999 9999999999999999 in the first instant of the Big Bang. If 0.01 was uncomfortably close to 1, this number was ridiculous. Dicke and plenty of other astrophysicists figured that this couldn't be a coincidence. Omega must really be 1, and we just hadn't found the rest of the matter yet.

In addition to this "flatness problem," cosmologists had to face what came to be known as the horizon problem. If you look straight up from the North Pole and straight down from the South Pole, the heavens look essentially identical. Some details will differ, of course—different constellations, individual galaxies in different positions, and such. Overall, though, things are the same. There are the same numbers of galaxies, on average, and they look the same, and they cluster in the same way. Even the microwave background is the same: 2.7° C above absolute zero, same intensity.

But for this to be true, opposite sides of the universe would have to have been in contact with each other at some time in the past. Consider the microwave background (which, after all, comes from the stuff the galaxies evolved out of). In order to be a true blackbody, which it clearly was, it must have been in thermal equilibrium when it was emitted. That is, any gross temperature differences would have to have equalized; but for one region to "know" another was hotter, and to respond by sharing the heat, those regions could be no farther away from each other than the distance light could travel. If they were, the expansion would have whisked them apart, and equalization would never have taken place. You might counter that the universe was tiny in the beginning, and it was. But light would still have taken a finite time to go from one place to the other, and the expansion didn't permit enough time. (The opposite sides of the visible universe, it should be emphasized, are moving apart faster than the speed of light. Einstein's famous rule that nothing can do this applies to objects moving through space. The expanding universe is a stretching of space. So even though what's now the visible universe was much smaller 15 billion years ago, the microwave light we see today has just reentered our horizon. It's taken 15 billion years to reach us. A year from now, the radiation from just beyond our view will show up. And so on.)

Astronomers and physicists started wrestling with these questions in all sorts of different ways. To solve the flatness problem, for example, they supposed that there might be a lot more matter in the universe than they'd found so far—matter that for some reason didn't shine like the stars or like the clouds of hot gas between them. In fact, observations had been hinting at the existence of such "dark matter" since the 1930s. At that time a cranky, brilliant Swiss astrophysicist named Fritz Zwicky, then at Caltech, had used redshift measurements of galaxies in the great cluster of galaxies in the constellation Coma, to gauge not their speed of recession from the Earth, but their speed relative to each other. Zwicky determined that they were whirling around each other, like a solar system whose basic units were galaxies, not planets. But when he compared the masses of the galaxies with their speed of revolution, it didn't make sense. The galaxies were orbiting so fast that the cluster should long since have flown apart. Zwicky deduced that there must be an enormous amount of invisible matter in the cluster, whose gravity was holding the whole thing together.

Nobody paid much attention, though. Zwicky's measurements were on the edge of what was possible; he might easily just be wrong. The reaction was much the same in the 1950s and 1960s, when Vera Rubin, of the Carnegie Institution, along with the husband-and-wife team of Geoffrey and Margaret Burbidge (who, like Hoyle, remained loyal to the Steady State) measured the spin rate of the M31, the Andromeda galaxy—the same one Edwin Hubble had used to prove that the spiral nebulae lie outside the Milky Way. They measured the redshift, not of Andromeda as a whole, but of its leading and trailing edge, and of several points in between, right into the center. They saw, to their astonishment, that it was rotating too fast. Where Zwicky had deduced that clusters of galaxies should be flying apart, Rubin and the others calculated that Andromeda should be disintegrating. Again, some sort of invisible dark matter was the best, though not a very plausible, explanation—especially since calculations suggested that whatever it was must add up to ten times the mass of Andromeda's visible matter.

But the evidence wouldn't go away. In the early 1970s, Jim Peebles and his Princeton colleague Jeremiah Ostriker published a paper that demonstrated that the shape of the Milky Way galaxy, and thus, by implication, all spiral galaxies, was unstable in a different sense. Their delicate spiral structure should have long ago collapsed into a sort of bar shape. In order to remain a spiral, the galaxy must be embedded in a huge, roughly spherical halo of invisible matter, like a butterfly inside a glass paperweight—again, about ten times as much dark matter as visible. Rubin went out and measured the rotations of other galaxies as well, and the excessive speeds showed up wherever she looked.

By the early 1980s, it was pretty clear that the dark matter was really there. But what was it? The most straightforward assumption was that it was ordinary matter that didn't shine significantly—very dim stars, say, or giant, free-flying planets. Unfortunately, that turned out to be extremely unlikely when a group of theorists at the University of Chicago, led by the late David Schramm, used the equations of nuclear physics to calculate how much baryonic matter—that is, matter made from ordinary protons, neutrons, and electrons—could possibly have been manufactured in the furnace of the Big Bang. Not enough, they determined, to account for the rotations of galaxies. The dark matter must be something else—clouds of nonbaryonic elementary particles, perhaps, that were incapable of experiencing or generating electromagnetic radiation and thus couldn't shine.

For a while, the leading candidate was the neutrino, a particle that had long been presumed, though not proven, to be utterly massless. If it did have mass, maybe neutrinos could serve as the dark matter. But when theorists ran neutrinos through computer simulations, they made a universe where galaxies formed very late—and observations had already proven that galaxies formed early, just a few billion years after the Big Bang.

Ultimately, the theorists settled on a class of particles known as "cold" dark matter. Whereas neutrinos are "hot," which means they naturally move at near the speed of light, cold particles lumber along more slowly. Nobody had proven (nor have they proven to this day) that cold dark matter actually exists, but it did have several things going for it. First, Grand Unified Theories (GUTs), the most widely accepted theories of particle physics at the time, independently predicted the existence of such particles. Second, computer simulations that assumed their existence produced a universe that more or less resembled the real one. By the mid-1980s, the Standard Model of cosmology included both a Big Bang and cold dark matter. And although he was still just in graduate school, a promising young astrophysicist named David Spergel became deeply involved in the attempt to understand what these particles actually were. His Ph.D. thesis at Harvard suggested a strategy for detecting them: if the dark matter surrounding the Milky Way really was a cloud of cold particles, Earth should feel a "wind" as it orbited the Sun. When Earth's orbit was moving in the same direction as the Sun's motion around the galaxy, we'd be going with the flow; six months later, we should be going upstream, and thus feel a much stronger particle wind. (Unfortunately, while the concept was sound in theory, nobody's ever detected a dark-matter particle, so it hasn't been tested. They're still trying.)

While Spergel was still an undergraduate at Princeton, the Standard Model had acquired yet another complication, in the form of a bizarre idea known as inflation. Like cold dark matter and predictions of the existence of the CMB, inflation came not from astronomers but from elementary-particle physicists. The general ideas that led to inflation had been percolating in the particle-physics community for several years, and, like television, it arguably had multiple parents. The scientist generally credited with bringing those ideas together, though, is Alan Guth, who had passed through Princeton during graduate school and wound up at the Stanford Linear Accelerator Center in California. Guth wasn't interested in astrophysics at all: he was trying to find an explanation for why physicists have never discovered magnetic monopoles. Since electricity and magnetism are two aspects of the same force, and since electric charge can be isolated—electrons have negative charge only, protons have just positive—you'd think you could isolate the north and south "charge" of a magnet as well. But you can't: although theory suggests particles with just north or just south charges should exist, exhaustive searches have never turned up a single one.

Guth wanted to figure out why and was looking at the same Grand Unified Theories that gave rise to cold dark matter particles. These GUTs also implied that as the universe cooled in the first few seconds of its existence, the basic forces of the universe—electromagnetism, the strong nuclear force, gravity, and the weak nuclear force—became separate, one by one, where they had originally been unified. Each of these separations involved a phase transition, analogous to the freezing of water into ice, accompanied by changes in the underlying energy structure of empty space. What Guth found was that under some circumstances, these changes might be delayed—just as pure, undisturbed water can be "supercooled" below 0°C for a little while before it freezes. If the universe was supercooled this way, the delayed phase transition would release an incredible amount of energy, all at once. And as Guth worked long into the night of December 6, 1979, he realized that this energy would have the effect of turbocharging the expansion of the universe. During an inconceivably short period of time, as the visible universe went from being 10^{-35} of a second to 10^{-34} of a second old, it would have expanded from the size of a proton to the size of a grapefruit—a small distance in absolute terms, but immense compared with how long it took.

To Guth's great satisfaction, this rapid expansion would have diluted the concentration of magnetic monopoles to the point where one, at most, should remain in the visible universe. No wonder nobody had found it. But as he thought about it more, he realized that the inflationary episode would also have implications for the flatness problem he'd once heard Bob Dicke describe at a talk at Cornell, where Guth was a postdoctoral fellow. "SPECTACULAR REALIZATION," he wrote on a sheet of paper. "This kind of supercooling can explain why the universe today is so incredibly flat — and therefore resolve the fine-tuning paradox pointed out by Bob Dicke." His reasoning: even if the universe is curved overall, inflation would have stretched our portion of it to the point where the curvature is, to all intents and purposes, gone. Our visible universe started out as a small patch within the greater, curved universe just as an acre of land in Iowa is part of the curved Earth—and, just like that acre, it seems to be utterly flat. Where astronomers and physicists had invoked absolute flatness out of philosophical prejudice—the universe just couldn't be this close to flat and not all the way there—inflation gave a reason why it should be.

That's not all. Inflation, it turned out, also neatly solved the horizon problem. Opposite sides of the universe were in causal contact with each other, after all; then inflation kicked in and drove them apart, still bearing the heat and structural signature of their early intimacy. And finally, it offered the first plausible explanation for where the original density fluctuations, which led to the modern universe, had come from. They originated with the energy dump that triggered inflationary expansion. Quantum mechanics says that on very fine scales, no distribution of energy can be perfectly smooth. The so-called inflaton field that Guth described would have been foamy, and as the energy decayed into particles of matter, that foaminess would have been preserved, and, as inflation proceeded, maintained. What had been microscopic fluctuations in energy density were transformed into macroscopic fluctuations in matter density.

Guth's original inflationary scenario, it turned out, had some technical flaws, but these were fixed up by other theorists, including Paul Steinhardt and Andrei Linde. The only real problem with the theory was that there was absolutely no evidence to support it. Nevertheless, theorists loved it. "Inflation could turn out to be wrong," admitted particle theorist Joel Primack in a 1991 interview. "But no theory as beautiful as this has ever been wrong before." So in addition to cold dark matter and a hot Big Bang, the Standard Model of cosmology circa 1985 or so included inflation as well.

Observers, meanwhile, were hardly content to sit patiently by while the theorists played with their equations. A complete model of the universe had to explain how it reached its present form. As of the late 1970s, nobody really had a good idea of what that form actually was. Observers knew that many if not most galaxies were organized into clusters—the Coma Cluster, which Zwicky had examined; the Virgo cluster, etc. On a large scale, the clusters themselves were organized into superclusters. But nobody knew the extent of this clustering—how big the largest structures in the universe were. They assumed that at some point any texture in the universe would smooth out, much as the dots in a pointillist painting merge into a smooth image when seen from far enough away. They also assumed that the superclusters were the biggest things around, and that they, along with ordinary clusters and individual galaxies, were spread pretty evenly through space. They had to assume, because the actual distances to galaxies, which you'd need for constructing an accurate three-dimensional map of their distribution, hadn't been measured—no more than a few hundred of them, anyway.

In 1979, though, while on a search for unusually faint galaxies, Robert Kirshner, Augustus Oemler, and Paul Schechter did a concentrated bout of distance measurements and discovered to their shock that galaxies weren't evenly spread at all. In the direction of the constellation Bootes, at least, there was evidently a great void, a region several hundred million light-years across, with almost no galaxies at all inside. (When their discovery was announced, the *National Enquirer* ran the headline "DID A REAL-LIFE STAR WARS CREATE HUGE HOLE IN SPACE?")

The Great Void in Bootes could have been a statistical fluke, but a careful survey of galaxy distances by Harvard astronomers John Huchra and Margaret Geller proved over the next several years that it was not. In 1986, they and graduate student Valerie de Lapparent released the redshifts of some 15,000 galaxies. Voids the size of what Kirshner et al. had found were clearly common, with galaxies confined mostly to the edges. The modern-day universe was more highly structured than anyone had suspected, which meant the early universe must have been as well.

The assumption that there was some kind of structure in the cosmic microwave background had been around essentially since it had been discovered in 1965. For one thing, the Earth isn't stationary; it orbits the Sun, which orbits the center of the Milky Way. All of this collective motion adds up to a velocity of about 200 kilometers per second. The CMB, though, propagates with equal intensity in all directions. So like the "aether," the stationary medium through which nineteenth-century physicists (falsely) believed all motion takes place and electromagnetic radiation propagates, it can act as a reference frame against which all other motion can be measured. The Earth's collective motion, therefore, should show up as a slight blueshift in the CMB in one direction and an equal amount of redshifting in the opposite direction. The British cosmologist Dennis Sciama and Wilkinson's original partner, Jim Peebles, both suggested that this wavelength-shifting should be detectable. (Peebles described this notion in his 1971 book, *Physical Cosmology*, in a chapter whimsically titled "The Aether Drift Experiment." This was a reference to Albert Michelson and Edward Morley's celebrated 1887 experiment, based on an 1881 experiment by Michelson alone, which tried to demonstrate the existence of the original aether. If Earth was indeed swimming through the invisible, undetectable aether, then a light beam moving parallel to the planet's path through space should travel at a different speed from one moving at right angles to our motion; Michelson and Morley fully expected to prove its existence. But when they measured the velocity of light both in and at right angles to the Earth's motion, the speeds were identical. There was clearly no aether.)

In modern terms, the CMB itself should be completely uniform, and the Earth's motion should create the illusion of a bidirectional, or dipolar, imbalance. So in the late 1960s David Wilkinson and Princeton colleague Bruce Partridge (now at Haverford) set out, first of all, to see whether the CMB was generally uniform, as the Big Bang demanded, and then to detect this slight departure from uniformity. "We didn't find it," Wilkinson said, "but I looked back recently to see how close we were; our error bars were only about a factor of two too big." He and a series of graduate students and other colleagues tried again later, using a balloon to lift the instruments away from the ground and above some of the atmosphere. This time, he said, "we saw a signal, and wrote down what that would mean for a dipole *if* it was a dipole. But we were pretty conservative, and decided we would not announce the discovery. Looking back, it was pretty clear that we were seeing it—but not with enough confidence that you would step up and toot your horn." It was also clear that this dipole was pointing in the wrong direction. It wasn't oriented along the Earth's path around the Milky Way, but instead was pointed in almost the opposite direction, toward the Hydra-Centaurus supercluster of galaxies, and it was moving much too fast. "It was as if," he said, "the whole Milky Way was being pulled in that direction."

In 1977, another team of astrophysicists, led by George Smoot at the Lawrence Berkeley Laboratory, did toot its horn. Flying a microwave detector aboard a U-2 spy plane, Smoot and his colleagues, Marc Gorenstein and Richard Mueller, had detected the dipole with high confidence. Smoot was clearly a more competitive fellow than Wilkinson. "Our competitors," he wrote in his book *Wrinkles in Time*, "particularly the Princeton team of Brian Corey and David Wilkinson, were hot on the track of the dipole. . . . It was important to us to get our work on the record." Unlike the original microwave experiments, this one didn't use a cold load as a reference against which to measure the cosmic signal. It was a differential receiver, which switched back and forth between two points in the sky and measured the difference in temperature between them. The CMB was its own reference point. Smoot had not only confirmed Wilkinson's earlier detections, but had also been able to show that the universe is not rotating (it might have been, after all, and if it was, there would be a distortion in the CMB as a result). Nor, he determined, was there any distortion suggesting that it was expanding in a lopsided fashion, as some theorists imagined might be the case.

Like Wilkinson's group, Smoot also saw that the dipole was pointing the wrong way. The Milky Way and all of its nearby neighbor galaxies were being pulled en masse at some 600 kilometers per second toward the Hydra-Centaurus supercluster. Bob Kirshner's "great void" had suggested that the universe was surprisingly lumpy in structure; this discovery, which implied a big concentration of mass and its associated gravity, suggested the same thing—as would observations by Geller and Huchra and by a group of astronomers affectionately known as the "Seven Samurai" during the next few years.

But Smoot and Wilkinson were both involved with another crucial CMB experiment by now as well. Back in 1974, it was clear to everyone who was interested in the CMB that cosmic microwaves could be observed most easily from space, as far away as from extraneous sources of microwave noise as possible. Three different teams of observers had applied to NASA with formal proposals to do so. Smoot was part of one group, based in Berkeley and headed by the distinguished physicist Luis Alvarez; Dave Wilkinson was on a second, which also included John Mather of the Goddard Institute for Space Sciences and Rainer Weiss of MIT; the third team came out of the Jet Propulsion Laboratory in Pasadena. Rather than simply choose among the three, NASA ultimately decided that all of the teams should join forces and build a single satellite.

As it took shape in 1976 and 1977, the space probe, the Cosmic Background Explorer, or COBE, would consist of three separate experiments, with John Mather in charge of the satellite as a whole. The Diffuse Infrared Radiation Background Experiment, or DIRBE, would look for the highly redshifted light that had been emitted by the very first generation of stars to turn on after the Big Bang; Mike Hauser, of the Goddard Space Flight Center, in Greenbelt, Maryland, would be the principal investigator. The Far Infrared Absolute Spectrophotometer (FIRAS), under Mather's direction, would make an ultraprecise measurement of the CMB spectrum to determine once and for all whether it was a true blackbody. And the Differential Microwave Radiometer, or DMR, would look for fluctuations in the intensity of the CMB—the ripples that must be there if the Big Bang model was to make sense. David Wilkinson was the obvious choice to be principal investigator, but he wanted no part of it. Wilkinson liked doing small-scale, hands-on science, and he dreaded the idea of being swallowed by NASA's bureaucratic machinery. Alvarez had never intended to work on the satellite at all. The job eventually went to George Smoot.

In the end, it would take more than a decade to get COBE into space, partly because the design was so complicated and partly because the satellite was originally designed to be launched from the Space Shuttle. When the *Challenger* exploded soon after launch in early 1986, that plan was abruptly changed. COBE had to be reconfigured for launch aboard a Delta rocket instead. It finally went into orbit on November 18, 1989, fifteen years after Mather, Smoot, Wilkinson, and the others had begun thinking seriously about their separate CMB satellite missions. It had taken so long, in fact, that its mission had changed substantially. Originally, the primary goal of the DMR experiment had been to detect the aether drift, as described by Peebles. That had been done long since. But nobody had yet seen the smaller variations that had given rise to the modern universe.

According to the Standard Model circa 1975 or so, that was a serious problem. In order for initial density fluctuations to grow into clusters and superclusters of galaxies, they'd have to be well along by the time of decoupling—about a tenth of a percentage point higher or lower than average, with a corresponding difference in temperature. At that intensity, they should be visible even from ground-based experiments. Cold dark matter, though, had given the theorists a reprieve. Because it couldn't interact with photons (the particles that carry electromagnetic energy) or even with itself, it could clump without betraying its presence; CDM would remain cold and dark no matter how dense it was. If you believed the universe was made mostly of CDM, you could posit a lot of structure in the cosmic microwave background without actually having to see it.

At some level, however, the fluctuations had to show up. "The theorists," recalled Wilkinson, "had decided that we'd better see them at a level of few times ten to the minus fifth [i.e., about a hundred times fainter than the original predictions]. If you don't see it then, they said, we'll fall on our swords." So that was one important goal of COBE's fluctuation-measurement experiment. Another was to look for the relative intensities of fluctuations at different scales. Back in the 1970s, theorists had suggested that the fluctuations should depart from average by about the same amount whether you measured large or small patches of sky. They had no explanation for why this should be so, but it was the least arbitrary supposition. Then, when inflation came along, it provided an explanation. The inherent foaminess of spacetime should, according to the rules of quantum mechanics, be scale invariant, and that scale invariance should come along for the ride when the universe inflated. If COBE detected such a pattern, inflation would pass its first real-world test.

COBE's Differential Microwave Radiometer team wouldn't be ready to report the results of its complex and delicate measurements for nearly three years, but the FIRAS instrument, which was looking at the spectrum of the microwave background, had a relatively quick and easy job. It was also an urgent job, thanks to a disquieting experiment that had taken place a few years before. Up until then, the blackbody nature of the microwave spectrum was reasonably well established: first Penzias and Wilson, then Wilkinson and Roll, had measured individual points on that spectrum. Those points fell on the curve that theory demanded, within a reasonable margin of error. So had subsequent measurements at a handful of wavelengths.

Early in 1987, however, a group of observers from Berkeley and from Nagoya University, in Japan, had sent a rocket 200 miles into space from the Japanese island of Kyushu with its own microwave detector aboard. The instrument took data at six wavelengths before falling back to Earth. And these data points didn't fit the established theoretical predictions at all. There was a bump on the curve, an excess of up to 10 percent over the expected intensity of microwaves. Maybe it was just an instrument error. Maybe some gigantic release of energy soon after the Big Bang had superimposed its own signature on the original blackbody curve, and the Berkeley-Nagoya rocket was seeing both events at once. Or maybe the microwave background simply wasn't a blackbody. Maybe the 1978 Nobel Prize had been awarded to Penzias and Wilson for something they really hadn't done. Perhaps the Big Bang had never been confirmed after all.

On January 13, 1990, less than two months after COBE went into orbit, John Mather and his colleagues on the FIRAS experiment were ready to answer these questions. Their instrument was a thousand times more sensitive than the Berkeley-Nagoya rocket; it would sample dozens of wavelengths compared with the rocket's six. Mather stepped up to the podium at the American Astronomical Society's winter meeting outside of Washington. He'd been worried that his talk, coming at the very end of the conference, would be poorly attended. He was astonished to find, in fact, that the huge ballroom at the Marriott Hotel where the conference was being held was packed to overflowing. Evidently his colleagues thought this announcement was just as important as he did. The man who introduced him was, ironically, Geoffrey Burbidge, one of the few anti–Big Bang cosmologists left in the world. Mather described how the instrument operated—the traditional exercise in torture that speakers inflict on audiences burning to see an exciting result—and then put an image of the plotted data up on the screen. It was a blackbody curve so perfect it might have come from a physics textbook. The assembled astronomers were silent for a moment. Then they broke into wild applause and rose, as one, to their feet. "That never happens at scientific meetings," said Wilkinson. "But I'm not surprised. When I first saw that curve, a month or so before they released it, the hair literally stood up on the back of my neck. I'd been measuring that curve, point by agonizing point, for twenty-five years. And suddenly, there was the whole thing, just laid out before me."

COBE had dropped its first shoe. It would take more than two years before it would drop the second.

CHAPTER 4

Bad Blood

For the reporters who covered it and the readers and viewers who experienced it vicariously the next day, the press conference held on April 23, 1992, in a cavernous exhibition hall at the Ramada Renaissance Techworld Inn, a conference center in Washington, D.C., was a thrilling, historic moment. It was here, at the tail end of a meeting of the American Physical Society, that George Smoot declared he'd seen the face of God.

Sort of, anyway. What Smoot had actually seen emerging from the data beamed back to Earth from the COBE satellite was evidence at last—and to everyone's enormous relief—of fluctuations in the Cosmic Microwave Background radiation. The average temperature of the CMB was 2.7° C above absolute zero, just as Arno Penzias and Robert Wilson had said in 1964, and as dozens of other observers had continued to say ever since. But COBE was finally sensitive enough to show that the actual temperature varied by one ten-thousandth of a degree up or down, depending on where on the sky you pointed your micro-wave detectors. COBE was at last seeing evidence of very slight variations in matter distribution, already evident when the universe was just 300,000 years old, that would evolve over billions of years into huge patches of the universe, hundreds of millions of light-years across, with a few more galaxies than average, or a few less. How important was this, asked one reporter? Could he put it into language everyone could understand? "If you're religious," said Smoot, "it's like seeing God."

No reporter could let that sound bite go by. None did. Smoot's dramatic pronouncement was on every news broadcast that night, and on the front page of every major U.S. newspaper the next morning. Other scientists, asked to comment, tried to outdo Smoot with their own superlatives. The discovery was, said Mike Turner of the University of Chicago, "the Holy Grail of cosmology." (If you'd been keeping track, you'd have noticed that it was one of six or eight Holy Grails of cosmology proclaimed by scientists and science reporters over the previous decade, to say nothing of the half-dozen or so more claimed for physics. Holy Grails were a growth industry in the 1980s.) Joel Primack, at the University of California, Santa Cruz, predicted the detection would prove worthy of a Nobel Prize. And as if that weren't enough, Stephen Hawking declared that the COBE discovery was "the scientific discovery of the century, if not all time." Journalists who cover physics and astrophysics know that Turner can be counted on for the most colorful quotes on just about any topic. But praise from Hawking, the most celebrated scientist of any kind on the planet, is the Holy Grail of science journalism.

Largely as a result of these four remarks (God, Grail, Nobel, Hawking) Smoot—until that week almost completely unknown to the general public, and even to much of the astronomy-literate public—was transformed overnight into a celebrity. He showed up on essentially every TV talk show and morning program to explain his amazing discovery. Newspaper reporters begged for a few minutes of his time. Within a day, John Brockman, a literary agent infamous for turning scientific discoveries into hugely profitable book deals, had called from a pay phone in Japan to sign Smoot up. (The resulting proposal reportedly sold for $2 million, which is not at all implausible.)

What the public didn't know at the time was that plenty of other members of the COBE team were ready to murder George Smoot. One reason was that he didn't work very hard to explain that the satellite was a team effort. From most of the stories published and broadcast on the COBE discovery, you'd easily have gotten the impression that one scientist was responsible. In interviews, and eventually in his popular book *Wrinkles in Time*, Smoot did little to discourage the idea that COBE was a Smoot production. "He had slogged through all the early steps: the definition of the question and the development of a way to answer it, with a set of instruments to be flown on a satellite," wrote John Noble Wilford in a *New York Times* profile of Smoot that came out a few weeks after the discovery.

This wasn't entirely Smoot's fault. Journalists find it much easier to tell a story if there's an identifiable hero. Somebody had to be the focus of attention, and Smoot was the principal investigator, or PI—the head of the team—that measured the fluctuations in the microwave background. Nobody could reasonably expect him to explain patiently, in one interview after another, that his was just one of three experiments on COBE, that he was just one of many scientists responsible for the discovery—and just one of four who spoke at the initial press conference—and that, despite the Hawking imprimatur and the talk of Nobel prizes, what the Differential Microwave Radiometer experiment had found wasn't a huge surprise. The reporters weren't really interested in hearing any of that, anyway, and nobody with an ordinary ego could easily resist being made into a hero.

But some of his colleagues had begun to feel, especially during the last few months leading up to the Washington press conference, that George Smoot had more than an ordinary ego. There was, for example, his run-in with Ned Wright, a theorist from the University of California, Los Angeles. His colleagues consider Wright to be an unusually smart and creative scientist in a profession that has more than its share of intelligence and creativity. He's especially good at analyzing data, teasing useful information out of a mountain of noise. And in the summer of 1991, he had come up with an algorithmic shortcut—a clever bit of software—that let him comb the data pouring in from COBE faster than George Smoot's group at Lawrence Berkeley could.

That evidently bothered Smoot quite a bit. It wasn't that there was any question of Wright's going public with his analysis prematurely. For one thing, the team rules forbade such a move. For another, it would have been absurdly reckless. By the end of the summer, he was pretty sure he saw evidence of fluctuations, but just about everyone agreed that without a detailed analysis of the possible sources of error—which hadn't been done yet—the evidence had to be highly suspect. Still, Smoot appeared to be threatened by Wright's independence. That, reasoned Chuck Bennett, of the Goddard Space Flight Center, Smoot's deputy principal investigator, was the only possible motivation for Smoot's declaration that "I don't want Ned to have access to my data." But it wasn't his data; the data belonged to the collaboration, not to Smoot. (Smoot himself says he prefers not to rehash ancient animosities; the important thing, he says, is the fantastic science done by COBE. But he does say, while declining to name names or address specific accusations, that "some people involved in these events are not telling the truth.")

Bennett was also somewhat appalled when Smoot proposed doling out to the team data that had been deliberately contaminated with fake signals. The most charitable interpretation would be to accept Smoot's stated motivation: just as with a double-blind medical test, a "placebo" set of data would assure that the analysis was free of bias. He'd have the right data, and everyone else would have the doctored set. "George asked me to present this idea to a meeting of the Science Working Group that he couldn't attend," says Bennett. So I got up and presented his idea like a good soldier." After he was done, there was a moment of silence. Then Ray Weiss of MIT, head of the working group, said, "Are you crazy?" Bennett answered that it wasn't his idea. "OK, then, sit down and let's get back to work," said Weiss.

In the end, the group decided to split up into small teams and reanalyze the data, some using Wright's software, some with the official software. In November, meanwhile, Smoot headed off to Antarctica to work on his own ground-based CMB detection experiment, leaving an annoyed Bennett to act as Principal Investigator himself, just as the project was reaching its climax. By the time he returned a few weeks later, Bennett, Wright, and COBE team member Al Kogut had written up their results in three separate papers, including Kogut's exhaustive analysis of the sources of error. They thought it was nearly time to go public, but Bennett remembers that Smoot was "furious" at this suggestion. As John Mather and journalist John Boslough report in their book *The Very First Light*, Bennett was of the opinion that

"George's outburst was clearly an act of intimidation. He was upset that I had had the temerity to be the lead author on a paper using what he perceived as 'his data.'"

According to Bennett, Smoot's behavior wasn't all that surprising by this point. "When I first began working as deputy PI," he says, "people who had worked with him would ask me, 'How's that going, anyway?,' as if it might not necessarily be going so well. I didn't know what they meant at the time." It became apparent after a while, though. As deputy PI, Bennett was the highest-ranking DMR team member at the Goddard Space Flight Center, where COBE was being built and tested. He dealt with the day-to-day problems of putting the instrument and satellite together. He was also fresh out of graduate school at MIT, and while he felt tremendously lucky to fall into such an important role on such an important project, he was also overwhelmed. "It was a baptism by fire," he says. "It took me a while to understand how it all worked. George didn't spend a lot of time mentoring me—he just did his thing, and I'd struggle to make sure things kept moving."

Smoot's failure to give Bennett much guidance isn't so surprising considering that hiring Bennett wasn't Smoot's idea, it was Mike Hauser's and John Mather's. The reason they wanted Bennett, according to one prominent member of the COBE team, was that "everyone agreed that the experiment was in big trouble. John and Mike Hauser brought Chuck in to straighten it out." Aside from his professional credentials, Bennett already had a fairly impressive network of coincidental connections to cosmology research. His thesis advisor at MIT was Bernie Burke, who had put the Princeton and Bell Labs groups in contact with each other back in 1965. He had worked under Ken Turner, who told Burke about Peebles's talk on the topic that same year. He'd worked with the group headed by Vera Rubin, who was one of the early observers of dark matter. Rubin's parents were friends of the family—and, finally, Bennett's father had worked at the Aberdeen Proving Ground, where Edwin Hubble had spent time during World War I.

When Smoot did come to town to check on the progress of the satellite and meet with team members, which happened fairly often, Bennett recalls that "he'd typically blow up at the engineers, rake them over the coals for one thing or another. I could see how counterproductive that was, whether or not he was right or wrong on a particular issue. Often, when he would leave, some of the better engineers would come to me and say, 'I'm off this project.' Of course, they could afford to do it, since unlike the mediocre engineers, they could always get assigned to another project. The hardest part of doing a space mission isn't the technical part—it's all the personalities involved. It's important to be motivational. The best people work on projects because they care about them. If you get them excited about the project, you get people who can't wait to come in and work on their shift. I felt like I was the class psychiatrist. I'd say 'relax, it's just him, you're doing a great job'—which they were. Every time he'd come to town it was, like, tensing up to deal with the barrage."

Bennett noticed other kinds of what he calls "funniness" as well. "My favorite," he remembers, "is that at one point where Smoot came into town for one of the science team meetings. Some of this stuff was new to me—I did radio astronomy before—and every now and they I'd sit down and work through something to make sure I understood it. And I was working through one of the sets of equations on the dipole, and I realized for the first time, though I assumed everyone else knew this, that the dipole was only the leading term of an expansion, and that there was a next-order term there, and while people generally ignored it, it might be measurable by COBE."

When he told Smoot what he'd figured out, says Bennett, "My memory is that he said, 'Oh, no, that's not true.' Then the next day in the science team meeting, he got up and reported that he had figured this out. I'm sitting right there, thinking 'this is really weird. Does he not remember that I'm sitting here? Does he think I don't care?' I didn't say anything. It wasn't a big deal, but I wondered why, since it was just twelve hours since I'd brought it to his attention, was he pretending otherwise. Actually, I've come to believe he wasn't pretending—that he believed he *had* figured it out. It was clear, at least in retrospect, that other team members had had these incidents, because they kept asking me with great sympathy, 'How's it going dealing with George?'"

If Smoot wanted to hold back on releasing the DMR results because he was worried others would get too much credit, he was also acutely aware of how important it is to be absolutely sure of what you're seeing. But waiting too long meant he'd run the risk of being scooped—and the DMR team knew that other scientists operating from the ground were working hard on getting at least some of the answers before they did. By early January, everyone agreed they had a solid discovery. "The reaction at NASA headquarters," says Weiss, "was 'Wow!'" Everyone agreed that the DMR result would be announced on Thursday, April 23, at the annual meeting of the American Physical Society. That being the case, other team members were surprised and upset as the meeting approached to learn through third parties that Smoot had tried on his own to schedule a talk for Monday the 20th. It looked as though Smoot, who had been furious that his collaborators were trying to get undue credit, was now trying to grab credit for himself. "That's simply not true," says Smoot. "I did try to get a talk scheduled during the plenary session on Monday, when I knew the audience would be very large. I was trying to grab credit for COBE, not for myself." (He kept what he was doing secret from the COBE team, however.)

But then, on Tuesday, John Mather was shocked to receive a call from an Associated Press reporter who wanted an interview on the new DMR results. Those results were supposed to be a strict secret until Thursday—so how did he know? As Mather looked into the leak, he learned that it had come, not from a loose-lipped colleague whispering in the back corridors of the conference, but from a full-fledged press release issued by the Lawrence Berkeley Laboratory, Smoot's home institution. The release warned reporters not to print anything before Thursday. But because it hadn't been seen in advance or approved by the COBE Science Working Group, it violated the team policy. The release, says Mather, also gave the impression that COBE was a Lawrence Berkeley project, and mentioned only George Smoot by name. "I called George," says Weiss, "and asked, 'What the hell is going on?' Then I called his boss, and he had the temerity to tell me 'this is the proper thing to do,' and implied that nobody but George was qualified to comment on the data, anyway."

Smoot himself takes the blame for this lapse. "The director of my lab had had bad experiences with NASA before, where the agency didn't mention our lab at all. So he insisted we do our own release. If I'd paid closer attention, I would have realized this was against group policy. It was a big failure on my part." It was in the Berkeley press release that Stephen Hawking made his remark about "the scientific discovery of the century, if not all time." Coupled with Smoot's declaration that he'd seen the face of God, it turned an important scientific result into a media circus, and, for Smoot, a highly profitable one.

"The press conference itself was a charade," says Weiss. "Everybody was furious with George." Among the sanctions they considered were asking Smoot to resign from the collaboration, or replacing him as head of the DMR science team. At one point, Bennett was asked if he wanted to take on the job. Bennett said no. "George had worked really hard, and he's a really smart guy, and I just didn't feel that that would be appropriate." Weiss didn't push it.

"I thought it was stupid and wrong to air a fight over credit in public," says Weiss, and ultimately the SWG decided that it would be best to settle for a letter of apology from Smoot. Many of the team members felt that the apology wasn't entirely heartfelt. "Some people have accused George of being a lousy scientist," says Weiss, "and that's unfair. He worked hard for almost twenty years on this project, and he did a good job. I don't know what happened at the end. I did everything I could to help George. But I feel that he double-crossed me."

CHAPTER 5

Now What?

Personal animosities aside, COBE had been a spectacular success. The hot Big Bang theory, first sketched out in 1927 by Georges Lemaître, required an echo of microwave radiation with a blackbody spectrum; by 1992, there was already strong evidence that cosmic microwaves had such a spectrum, but COBE had delineated it with such remarkable precision that the CMB curve could have replaced theoretical versions in physics textbooks. No reasonable person could doubt the existence of the microwaves predicted in 1948 by Ralph Alpher and George Herman.

And now, to the relief of theorists, COBE had also found the seeds of cosmic structure. The over- and underdensities of matter represented by these temperature variations were minuscule, at just one part in one hundred thousand. They were so small that if ordinary baryonic matter, made of protons and neutrons and electrons, were all that was involved, the universe would still be much too young for gravity to have congealed them into stars and galaxies and superclusters. Before cold dark matter, or

CDM, that would have been a problem, but its existence was confirmed everywhere the observers looked. If CDM really existed, as most theorists believed, then its gravity would have jump-started the formation of structure without leaving any imprint on the microwave background. Perfect.

Not only that: the intensity of the fluctuations was, within the bounds of experimental error, reasonably close to being scale-invariant. That is to say, the amplitude, or intensity, of the fluctuations was about the same no matter how big the patches of sky you were comparing. In the cosmic ocean, there were majestic swells, ordinary waves, and tiny ripples, all superimposed on one another, but all reaching about the same height and plunging to the same depth. Scale-invariance was just what inflation theory predicted, so COBE was an enormous triumph for Guth and Steinhardt and Linde. (Stephen Hawking would later explain that this was why he'd declared the DMR result the "greatest discovery . . . of all time": it supported inflation.) It didn't prove that inflation was correct by a long shot, of course: consistency with a theory isn't the same as confirmation. But the DMR could have ruled inflation out, and it did not.

The COBE DMR result did, however, rule out a competing model for the growth of cosmic structure. According to a theory that had been advanced by Neil Turok, then at Princeton, and David Spergel, who had returned there after grad school and a stint at the Institute for Advanced Study, the seeds that grew into clusters and superclusters of galaxies were not fluctuations in the density of matter but "defects" in the topology of spacetime itself, created as the universe made imperfect transitions from one energy state to another. Resembling the lines and sheets of discontinuity that mark the spots where two crystals with different orientations met during formation, these defects would have trapped energy and gathered mass around them. But the pattern COBE found didn't fit this idea, and Spergel declared to reporters at the Washington meeting, without evident sour grapes or even regret, but with a smile and a tone of cheerful resignation: "We're dead."

David Spergel was at least two scientific generations younger than David Wilkinson, depending on how you calculate such things. When the latter David had been constructing his first CMB detector on the roof of Guyot Hall back in 1964, Spergel was only three years old. His father was a physicist, and would eventually become the head of science at York College, part of the City College of New York system. But while he exposed his kids to physics, he didn't try and push it as a career. As David went through school, though, it became clear that he was good at math and science in general, and interested in astronomy in particular. He did a Westinghouse Science Search project on understanding the rings of Uranus, for example, and when he arrived at Princeton as an undergraduate he was pretty sure he wanted to major in physics (though the Woodrow Wilson School of Public and International Affairs was a close second).

Physics won, and as Wilkinson had done two decades earlier, Spergel slid quickly over into astrophysics. "My junior paper was a detailed calculation of galactic dynamics under James Binney," he says. "It was a real project, not just busywork, and I really got caught up in it." When he went on to grad school at Harvard, Spergel hoped to work on another complex, messy calculation: the physics of fluctuations in the early universe. But his advisor felt there wasn't much useful work to do in that area at the time; Spergel got involved in dark matter instead.

COBE did leave some significant loose ends. Comparisons of CMB anisotropies with the clumpiness of the modern universe—

great voids surrounded by walls and sheets and filaments of galaxies—suggested that there cold dark matter alone couldn't explain everything about cosmic structure. Maybe neutrinos, the original "hot dark matter," had a little bit of mass after all—a bit of extra flavoring that could season the recipe for the cosmos just right. It also wasn't entirely clear how to get from the observed value of omega—about 0.3, if you added up all the visible matter and all of the dark matter whose effects you could see—to the value of 1.0 that theorists and inflation still required. Even so, the Standard Model as it came together in the 1980s was now looking solid: the universe started in a hot Big Bang; inflation jump-started the expansion, and inflation provided the seeds of cosmic structure.

But it was also clear that the cosmic microwave background had to be probed in much greater detail. Inflation hadn't been refuted, but a scale-invariant spectrum of fluctuations had been a generic prediction of many models; finding it didn't prove inflation. Features smaller than seven degrees across were too fine for COBE to see at all, so you couldn't be sure that the inflation-friendly spectrum of anisotropies the DMR detected continued down to small scales. It was on these same small scales that all the action of structure building had begun. COBE, moreover, had flown with technology that was more than a decade old, thanks in part to the very long development time NASA had allowed for it and in part to the *Challenger* crash. With the more sensitive radiometers that had become available by the early 1990s, it was obvious that a new CMB experiment could drive down the margins of error in COBE's measurements, and nail down the physical characteristics of the early universe with a lot higher precision.

For David Spergel, that understanding came as something of a revelation. At Princeton, research on the cosmic microwave background, both experimental and theoretical, had traditionally been run out of the physics department, where it started in the 1960s. Dicke, Wilkinson, Peebles, and Roll had all been physicists; Paul Steinhardt, one of inflation's founding fathers, had joined the department from the University of Pennsylvania in the 1990s. But Spergel was on the faculty of the Department of Astrophysical Sciences, located in Peyton Hall, a few hundred feet uphill from the physics building. (The name "Astrophysical Sciences" sounds a little pretentious, considering that other high-powered universities leave it at "Astronomy" or "Astrophysics." It was evidently a political ploy by Lyman Spitzer, who headed the department in the 1950s and '60s and who is credited with the original idea for both the Hubble Space Telescope and for trying to produce power through controlled nuclear fusion. In order to study the universe in all its complexity, Spitzer wanted to hire physicists and mathematicians as well as straight astronomers. The math and physics departments objected to Spitzer poaching in their fields of expertise. So he changed the name of his department, and thus expanded his mandate.)

For Spergel, cosmic textures had been the next stop after his work on detecting dark matter. But while he admits he was a little depressed on having his pet theory demolished, that didn't discourage his interest; instead, it whetted his appetite. "Right after the COBE announcement," he says, "everyone was very excited and we all wanted to discuss what it meant. So I decided we should have a workshop. I put one together, here in the auditorium. I got together with Dave Wilkinson, and Lyman, Page, and people from this department and the Institute for Advanced

Study. And because we were doing this on short notice, we agreed, 'Why doesn't everybody try to invite one or two people? We can put them up at our houses. That's the easiest way to put together a conference on one month's notice.' We put together a list of names, and as we were going through the names, we wanted to make sure we invited all the relevant people from the COBE team. Smoot and Mather were invited [Smoot, perhaps not surprisingly, stayed away] and Dave said, 'We should invite Chuck Bennett.' I didn't know him very well, but he was one of the people who really did a lot of work on COBE."

The meeting was a great success. What Spergel remembers best is a talk by Dick Bond, a theorist from the Canadian Institute for Theoretical Astrophysics, a round-faced, bland-looking scientist who operates on such a rarefied level that graduate students tend to come out of his office dazed, their brains reeling from having tried to keep up with him. Bond argued that finding large-scale ripples in the CMB was all very good, but that smaller fluctuations would be an observational gold mine; it would let cosmologists measure, directly, all sorts of cosmological parameters, including the expansion rate, geometry, matter density, matter-to-dark-matter ratio, and other basic facts about the universe that observers had been struggling for more than a decade—and in some cases more than half a century—to nail down.

This was familiar territory for Bond. During the 1980s, he and a small handful of other theorists, including Jim Peebles, had been thinking about what you could learn from the CMB, and they had published several papers describing in detail what the small-scale ripples ought to look like, depending on what sort of physics was operating shortly after the Big Bang. On scales of a degree or less on the sky—about twice the size of the full Moon—they argued that the ripples should look fundamentally different from those on larger scales. These small ripples had, like the big ones COBE saw, started out with an equal degree of scale-invariance. When both sets of ripples were generated, presumably during the inflationary epoch when the universe was 0.000000000000000000000000000000001 second old, they had different wavelengths but very similar amplitudes.

As the universe expanded, the ripples began to respond to local physics. Regions of especially high density would act like tightly compressed springs: under the intense pressure of energetic photons, they'd try to expand into areas of low density. As the pressure was relieved in one spot, it would build in another, creating a new high-density region that would tend to expand in turn. For the first 300,000 years of its existence, until the photons could finally fly unimpeded through space, the universe would be percolating with ripples of density of all sizes, moving in all directions. When pressure waves move through air and enter our ears, we call them sound waves—but to a physicist, a sound wave is any wave of pressure that moves through any medium, whether it's a liquid, a solid, or a gas. The waves of pressure that propagate through the Earth after an earthquake are sound waves as well. So are the waves of pressure that ripple through the oceans: whales emit and listen to sounds that don't involve air at all.

These acoustic waves move at different speeds, depending on the density of the medium, so the speed of sound in rock is higher than the speed of sound in water, which is higher than the speed of sound in air (which itself can be faster or slower depending on temperature: sound travels faster on a freezing cold day, when the air is denser, than on a hot day). In the early universe, the density of matter was so high that the speed of sound was very nearly the speed of light—the ultimate speed limit, according to special relativity, of anything moving through the universe.

That being the case, no ripple could begin to evolve until it had time to become "aware" of its own existence. How long that was depended on how large the ripple was when it started out. A ripple ten light-years across couldn't begin to expand until the universe was ten years old. A ripple a thousand light-years across would take a thousand years to start moving. By the time the universe was 300,000 years old, any ripple 300,000 light-years across or less would have had time to start evolving. Anything larger would still be in its primordial state, frozen in place awaiting an awareness of its own existence—an awareness that would come too late for any record of its awakening to be left on the CMB.

By now, about 13 billion years after the Big Bang, regions that were 300,000 light-years across back then span just about a degree on the sky, which seems to make no sense; if that was the distance light could have traveled back then, then the entire visible universe today should be that same patch, grown larger. Once again, however, the universe can and does expand faster than light, even during a noninflationary period; distant regions that had expanded outside our horizon early on have gradually reentered as the expansion has slowed and as light has finally made its way back to us. Features in the CMB bigger than a degree across, therefore, and certainly those COBE saw, should still carry the pattern of fluctuations they were born with. Those smaller than a degree across should have had their birth pattern altered by the reverberations of acoustic waves that boomed across the universe. At the time of decoupling, when all patterns were fixed in the microwave background, some of the initial fluctuations would have had time to go through one full cycle of expansion and contraction; others would have gone through three or four or sixty or 5 million—and still others would be caught on the upswing or downswing, in between maximum compression and maximum rarefaction.

At any one point on the sky, the density of the universe at the time of recombination and decoupling, and thus the temperature of the microwave background, would be a complicated mix of all these different waves, caught at all different points in their boom-and-bust cycles, superimposed on each other. The result, said Bond and George Efstathiou and other theorists, should be not a scale-invariant spectrum of fluctuations, but a pattern with marked highs and lows at different angular scales—the highs coming where overlapping fluctuations of different length scales reinforced one another, the lows where they tended to cancel out. Because they were caused by sound waves, the highs were called "acoustic peaks." If you plotted the entire range of microwave background fluctuations, from the largest to the smallest scales, on a graph comparing angular scale versus intensity, you'd have a more or less flat line above the one-degree scale, showing the scale invariance of the unmodified initial fluctuations. Below it, you'd have a series of acoustic peaks, separated by valleys, and the height and location of the peaks should come in a predictable pattern, depending on the geometry and composition of the universe. By comparing the actual peaks with the theoretical predictions, you could presumably rule out or rule in a half-dozen basic cosmological parameters.

That was the idea, anyway. Before COBE proved that cosmic ripples actually existed, though, very few theorists invested much time in figuring out how the underlying cosmic parameters might affect the pattern of small-scale fluctuations—an exercise Spergel calls "parameter tickling." When COBE was under construction, in fact, Wilkinson went around to theorists and asked them how big the satellite's beams should be. "We were going to a lot of trouble making these antennas, after all," he said, "but everyone said 'doesn't matter.' So we ended up with these seven-degree beams." After COBE reported in, says Spergel, "all of that changed. Parameter tickling became a whole industry. And I decided that the CMB was where I wanted to be intellectually. After that meeting, I understood how powerful a tool it was, for two reasons. First, you could use it to make theoretical predictions, because the universe was really, really simple back then. And second, you could measure the fluctuations in the early universe with high precision."

CHAPTER $\boxed{6}$

Forming a Team

Exactly how to make these small-scale measurements was a matter of constant discussion in the cosmological communty. At Goddard, Chuck Bennett had been busy working on papers that refined the COBE results, but he was listening with one ear to the ideas that were floating around. "People were talking about ground experiments, balloon experiments, space experiments," he remembers. "I wasn't at all convinced that you needed a space mission. And I wasn't convinced you didn't, either. But you've got to really convince yourself that the stuff can't be done some other way, because space machines are expensive and complex—and frankly, I thought initially that the arguments for going back into space were weak."

For his part, Dave Wilkinson wasn't sure he wanted to work on a space mission again even if the arguments in favor of it turned out to be strong: COBE had turned into a nightmare for him. The Smoot problem was only a small part of it. A bigger source of frustration was the fact that the COBE team was very large, which meant that some members got away with doing very little work—not out of laziness, necessarily, but because they could get away with concentrating on other projects and still get credit for having been part of this high-profile mission. Another problem for Wilkinson was that COBE had been built almost entirely at Goddard. Scientists had done a lot of the initial design work and written up the COBE proposal, but after that it was turned over to engineers to carry out. "They were good engineers, obviously, because the thing worked," he said. "But I didn't feel that I was able to contribute much. I'm happier with a more hands-on style of working."

He understood why NASA had insisted on this approach. "They used to be in the mode where they had academics building satellites, but they found that college professors couldn't meet a schedule or stay on a budget, so they took over." But understanding it didn't mean he was willing to put up with it. Yet another frustration: some of the data analysis was handled not by the project scientists, but by outside contractors. In Wilkinson's opinion, they never performed very well. "They're just used to a different kind of job than thinking carefully about science data." If he was going to work on another space mission, it would have to be different from COBE in just about every conceivable way.

Just down the corridor from Wilkinson in Jadwin Hall, Lyman Page was also thinking about what to do next about the microwave background. Page, a tall, blond, earnest scientist who looks at least a decade younger than his four decades or so, hadn't worked on COBE at all. While the satellite was humming along in orbit, Page had been doing his doctoral dissertation as part of a team operating a competing, balloon-borne CMB experiment. (His advisor, Steve Meyer, was working on both; he had to be careful not to reveal the results of one experiment to members of the other.) Page was, by now, firmly a part of the microwave-background community, but in some sense it was accidental. Back as an undergraduate physics major at Bowdoin College in the early 1980s, he remembered being taken with some of Bob Dicke's ideas about the early universe—but not taken enough to plunge into a standard career. Instead, he took a job after graduation working in Antarctica, where he spent more than a year, including the frigid winter when the sun didn't rise for six months. His job was to tend other people's experiments. Living in an isolated hut several miles from the main U.S. Antarctic base at McMurdo Station, he spent much of his time reading, painting, and venturing outside to gaze in awe at the Southern Lights.

On his return to the United States, Page still didn't know what he wanted to do. After seventeen months in isolation, he still needed some time alone. So he bought an old sailboat in Maine and cruised down the East Coast. He was in the Caribbean, headed for the Panama Canal and from there on to the South Pacific, when a storm blew up and damaged the boat severely. That night, he decided that instead of Tahiti, he'd go to grad school in physics. He hadn't even taken the entrance exams, but he showed up on the MIT campus, where Ray Weiss looked up one day to see a bespectacled young man peering in with evident fascination at Weiss's cluttered lab in a prewar building that Weiss affectionately refers to as a shithouse. "I told him to come on in," says Weiss, who gave him a job as a technician, then lobbied to have him admitted as a grad student. "He didn't have a lot of credentials," says Weiss. "But he knew how to think on his feet, how to figure out how to make an instrument work."

Although Weiss was deeply involved in COBE at the time, Page didn't want to join him. He felt that he didn't have a lot to contribute to a project that was already overstaffed. Instead, he chose to work with Steve Meyer, who was launching CMB detectors from the National Scientific Balloon Facility at Palestine, Texas. In the months before Smoot declared he'd seen God, Meyer's group knew they were seeing something, too. The data weren't quite clean enough for them to feel confident about making their own announcement, however, and they weren't the only ones. "In our papers we said 'it's a possibility,' but we couldn't rule out that there was some funny atmospheric fluctuation or something." Once the COBE team went public, it was clear in retrospect that Page's group had been seeing the real thing too. "Certainly it was in our data first, or in our computer outputs first," he says, but he still thinks keeping it quiet was the right decision. "Especially with the discovery of something like that," he says, "you want to be certain you see it. So it was a possibility, but we couldn't defend it to rule out everything else." At the time of the COBE announcement, Page had also become involved with several other CMB experiments either in the planning stages or already in operation, including one in Saskatoon, in western Canada; and one at Cerro Toco, high up in the Chilean Andes.

But he also thought it would be interesting and fun to work on a microwave-background satellite that could refine and extend the COBE results. So when he ran into Dave Wilkinson in the corridor one day, and Wilkinson asked, "Do you think you'd be interested in getting involved with a CMB satellite," he didn't hesitate. "I said, 'Yeah, absolutely.'" For his part, Wilkinson had decided that whatever the arguments against working on another satellite, the scientific promise was too great not to do it. A ground-based experiment could measure the CMB at small angular scales, but it would have more trouble extracting the radiation's already weak signal from the microwaves constantly being emitted from the ground and the instruments themselves.

That wouldn't be so much of a problem if those confounding microwaves had about the same intensity in all directions; you could do a correction with relative ease in that case. But that's not the case for a ground-based experiment. Imagine, says Page, you're measuring temperatures inside a room. "You look at the ceiling tiles," he says, "and over a fraction of the width of one of those tiles, the radiation from two different points is going to be the same. But if you look over any appreciable size, ten degrees, say, you know the room is radically different. There are hot lights over there, and windows over here, and bookcases and a computer monitor and a heating vent. It's like that with a ground-based experiment. There are mountains, there are big clumps of atmosphere going by. So it's very hard to make measurements on large scales."

You can still make confident statements about what's going on in small patches of sky, of course—just as you can within a single ceiling tile. But even if you take careful measurements over, say, a two-degree field of view, you can't say anything about what the CMB is doing overall at two degrees, because you have just one patch, and nothing to compare it with. You can't say much about the one-degree scale either, since you have only four independent patches that size; the statistics are too meager to be meaningful. At half a degree, with sixteen patches you can begin to say something, but your confidence level won't be very high. As Chuck Bennett would say later, when he'd been convinced that a satellite was necessary after all: "COBE did a pretty good job at scales above seven degrees, and the ground-based experiments can get up to half a degree or so. But there's a big gap between a half and seven, and you're never going to get that from the ground." It's also important, he would say, to cover that gap with a single instrument; each CMB experiment has its own peculiarities, so the bigger range of angular scales you can probe with one telescope, the better you can see how the CMB is acting at all scales.

Page was also very knowledgeable about the electronics of CMB detection, but the real in-house expert on electronics—the instrument maven, he might have been called in a more Yiddish-literate department—was Norm Jarosik, one of the odder members, in a professional sense, of the Princeton physics department. His title is "research physicist," and while he does some teaching, that's not part of his job description. Neither is doing research. What he does, with a skill that people like Wilkinson say is close to genius, is build electronic devices that others can use for their own experiments. Most people knew someone like Jarosik in high school: they tend to be quiet, completely unfashionable, and endlessly fascinated with circuit diagrams and small, unidentifiable objects with wires coming out of them. People like Jarosik can sift for hours through the "$5 a pound" bin in an electronics surplus store.

But while that sort of interest typically leads to an engineering career (and indeed, Jarosik's father was an electrical engineer), his scope was broader. "We had this series of Time-Life Books on the various sciences—there was one on physics and one on chemistry, and so on. I just loved looking through those, seeing the kinds of things people were doing. In one of them, I remember, they had two pages of pi written out, you know a few hundred digits at a time, and that was a really impressive thing that somebody could do that."

So when he went through the University of Buffalo, in his hometown, Jarosik majored in physics, not engineering, and at the same institution, in grad school, he specialized in condensed-matter physics—the study of what goes on in solid materials and the field that gave rise to transistors and semiconductors (and, lately, high-temperature superconductors as well). He spent a couple of years as a postdoc at Bell Labs, and then came over to Princeton to do a second postdoc. "I was hired to convert a radiometer into a high-mobility transistor," he says. "But when that project was over, they needed me for something else. Each project led to something else, and they eventually realized I could be useful to have around permanently."

"When Lyman and Dave came over to see me," he says, "I thought it made a lot of sense to put together a space mission. So we sat down and dreamed about how we might do it. Our ideas were a little too simple, of course." But they put together what they called a "white paper"—it was too sketchy to be called a proposal—and gave the project the provisional name "Princeton Isotropy Experiment." (Dave Wilkinson would sometimes refer to it as "PIE in the Sky," and he wasn't just talking about where it would be located.)

They sent the idea off to NASA, but given the sketchy nature of the white paper and the fact that NASA was facing a huge budget crisis, there was essentially zero chance it would go far. It went nowhere. What they needed at the very least was to make a convincing case that they could build the kind of instrument that could make the measurements they wanted, and that it could be incorporated successfully into a satellite. Without that, they were like inventors who wanted to talk General Motors into creating a whole new car based on a new type of engine that hadn't even been designed yet. They needed to find partners who knew how to put together an entire space mission. So Dave Wilkinson contacted some people he knew at the Jet Propulsion

Laboratory in Pasadena. Unlike the Goddard Space Flight Center, which is a NASA facility, JPL is owned and operated by the California Institute of Technology, but it's funded by NASA, and essentially its entire task is to do space missions for the government. It also differs from Goddard in another respect: while Goddard specializes mostly in Earth-orbital missions—the Hubble Space Telescope, COBE, and all sorts of lesser-known satellites—JPL has built most of the country's planetary-exploration probes. The *Pioneer* and *Voyager* spacecraft, which flew by Jupiter, Saturn, Uranus, and Neptune, came out of JPL; the Galileo probe, which became a satellite of Jupiter and photographed that planet and its moons in exquisite detail, was a JPL project; the Mars probes that have alternately explored and crashed into the red planet (one, infamously, when someone failed to convert English units into metric when calculating the final trajectory) came from JPL.

But JPL has also been involved in orbital missions, and it was there that Wilkinson went first to explore the next step: primordial sructure investigation. "They had some COBE people out there," he remembered, " and we met and talked about what we wanted to do, and I laid out my requirements." He would collaborate as long as he could work on his own terms: the team had to be small, with each member taking on plenty of responsibility, and with the scientists involved at every stage of the mission. No handing things off to engineers who wouldn't let them near the hardware. No multiple layers of bureaucracy to insulate them from the instrument. In fact, insisted Wilkinson, "we had to build the instrument ourselves, at Princeton, and the project had to be a true collaboration, not a JPL project. They didn't want to do it that way." The Princeton-JPL satellite got a name—"Primordial Structure Investigation," which replaced the original Princeton Isotropy Experiment—but that was as far as things went. "We danced some with them," says Page, "but it sort of died away."

Down at Goddard, meanwhile, Chuck Bennett had gotten wind of Wilkinson's interest in doing another microwave background satellite. So in September 1993, about a year and a half after the COBE results were announced, he placed a call to Princeton—not because he was sold yet on the idea of a satellite, but because he respected Wilkinson and wanted to hear what he was thinking. "'I hear you're thinking about a mission,'" Wilkinson recalled him saying. "I said, 'Yeah.' I knew Chuck was a good guy from the COBE experience. So we started talking." They agreed that in order to know if a post-COBE satellite was even possible, they'd have to answer a number of preliminary questions. Could it be done from the ground (which was Bennett's lingering concern)? If they went into space, what orbit would they want? How would they handle propulsion? What launch vehicle would be available for doing this? And how cheaply could they do the whole thing?

"Dave was definitely of the school of wanting to do the simplest thing possible," Bennett says. "He's anything but an empire builder. He doesn't want to do the biggest-budget thing; he wants to do quite the opposite." That fit in perfectly with the NASA philosophy at the time: stung by the loss of the billion-dollar Mars Observer probe, and by the embarrassment of sending the $2 billion Hubble Space Telescope into orbit with a misshapen mirror—and by the congressional response, which was to slash the agency's budget—NASA administrator Dan Goldin was pushing to do things "better, faster, cheaper."

Bennett and Wilkinson liked what they heard during that first conversation, and Bennett volunteered to set up a meeting to begin exploring the idea of a collaboration further. On October 19, 1993, the core of what would become the MAP team met at the Goddard campus, about twenty miles from central Washington, to talk things over. Wilkinson brought Jarosik and Page along. The Goddard group was officially led by Mike Hauser, who had been Principal Investigator, or PI, for the infrared background instrument on COBE, and who was chief of the spaceflight center's Laboratory for Astronomy and Solar Physics; John Mather also took part, as did a senior engineer named Dave Skillman, "the kind of guy," says Bennett, "who has a high enough level knowledge of things that he could give kind of rough, off-the-cuff answers to various questions of what's easy and hard to do in space." Dave Wilkinson had plenty of such questions: "How would you do attitude control? How cheaply could you do the whole thing? What launch vehicle would be available for doing this?"

Following that meeting, Bennett recalls, "Dave seemed very pleased, because he had answers to all of his initial questions, and more. After that we just kept talking by e-mail and by phone and stuff, throwing out more questions and figuring out the answers. 'What do you need to know? What do I need to know?' We also talked in a lot more depth how you would manage a space mission, again with the idea of doing something simple, as simple as possible. How many people should be on a team, and who should they be? We hadn't committed ourselves to do this yet—just to formulate the questions and figure out the answers."

But they were also rapidly developing a powerful respect for each other. In Wilkinson, Bennett saw the precise opposite of Smoot: no-nonsense, straightforward, practical, and not in the least egotistical. He already had a sense of this from their encounters on COBE, but it really hit home when the time came a year or so later to formalize the collaboration. They also agreed that, aside from the Smoot problem, COBE had suffered from having too large a science team. "A lot of times," Bennett says, "people are interested in the early stages of a project when it's heady and exciting. And then when it starts to transition to doing lots of real work, eighty hours a week, you find that some people, especially if they have other research interests, start saying 'Well, I don't have time to do that.'" As a result, he says, the team has to add people to do the actual work—while keeping the original team members who still want to put in their two cents. "In fact, that's how I ended up working on COBE."

Before they could go ahead, they needed to make one important decision. Someone had to be the PI, and it was clear that either Bennett, who headed up the Goddard axis of the group, or Wilkinson, who was running the operation at Princeton, would get the job. "In many ways," says Bennett, "I considered that whichever way it was going to be, we'd be co-PIs on it. I'd never do something Dave thought was the wrong thing to do, unless we talked about it and fought it out, or whatever." Bennett thought Wilkinson might prefer not to be PI, since the job involves a lot of detailed budgeting, negotiating contracts with outside subcontractors, frequent reports to NASA headquarters, and a lot of other bureaucratic irritations. Wilkinson, he knew, was interested in the instruments and the science.

"At the point where we had to make that decision," says Bennett, "Mike Hauser was still involved, playing a sort of facilitating role but not planning to stay on the project, and I told him I was ready, willing and able to be PI. And so during a telephone conference with the three of us, Mike said, 'There obviously has to be a PI, and I think it should be Chuck.' And Dave said 'yes.'

And that was the end of the discussion." Ultimately, Bennett would invent a title for Wilkinson: "Instrument Scientist." He wanted to convey the actual role Wilkinson would play, which was to guide the science and troubleshoot. "I think it worked very well," says Bennett. "It enables him to float a little bit above the fray, look at what he's worried about—he calls or sends e-mail with advice or criticism or suggestions or whatever."

In Bennett, meanwhile, Wilkinson found someone who wasn't wedded to the bureaucratic rules that had made his own COBE experience a nightmare. Bennett remembers that at one of the early team meetings in Princeton, "Dave and Norm started pounding on me, saying, 'Are we going to be allowed to touch this hardware?' It was really funny, because here they are trying to make a case that they want to touch the hardware, and I'm thinking, 'That's exactly what I want too.' They're saying, 'We don't care what the quality assurance rules are, we don't care what the NASA bureaucracy says, we need to get up in there.' And all I said was: 'Yes.' I think they were right to be skeptical, because it wasn't a typical way to build missions. But I was completely committed to doing that. So I pushed all through the program to make sure we had those kinds of arrangements." "In a week," Wilkinson would later write in an e-mail to Bennett, "you and I made more progress toward understanding than PSI [the JPL project] did in one and a half years."

Still, the project wouldn't go much farther without encouragement—and some startup money—from NASA. So in February 1994, just a few months after his initial meeting with the Princeton scientists and before he and Wilkinson had approached Wright, Meyer, and the rest, Bennett approached some intermediate-level administrators at headquarters. At that point, the proposed satellite not only had no science team: it also had no name; no concept aside from the broad notion of improving on both COBE and on ground- and balloon-based measurements; and no design. NASA had signaled that it wasn't looking for unsolicited proposals. But Bennett—a savvy NASA insider by now, thanks to the trial-by-fire experience of shepherding the DMR experiment along—knew that anyone who wanted to do a mission like this had to get headquarters thinking about it as early as possible. "I called up, and it was basically, 'Hi, how are you? We're thinking about a cosmic background mission. We could use some study money. What do you think about that?' Their answer was, 'Love the idea, don't have much money here.'"

He did have a little from Goddard, though; the mission looked sufficiently appealing to the Center's administration that he was given some preliminary funding to explore the idea further. With that vote of confidence, Bennett and Wilkinson set out to put together a team that could mount a CMB mission. Both men made lists of people they thought should be on it, and discussed them. One key criterion was time: some highly qualified people were taken off the list because they were known to be busy with other projects. The PI and the IS wanted to avoid what they'd both seen on COBE: people who wanted to be associated with the project but only showed up occasionally, and then just to hang out and listen to what was going on. When Wilkinson and Bennett agreed on someone, they'd issue an invitation. "Nobody turned us down," says Bennett.

Page and Jarosik were on the list by default, of course. So was John Mather, although not for long: he would, in a matter of months, be tapped as the lead scientist for a new space telescope—the Next Generation Space Telescope—that would be a successor to the Hubble. Bennett also wanted Gary Hinshaw on the nascent team. Hinshaw had been hired back in 1990 out of a faculty job at Oberlin College to help analyze the COBE data, concentrating mostly on writing and streamlining the software that turned raw measurements into a coherent picture of the microwave sky. "As Chuck and I saw it," says Hinshaw, "the talent I was going to bring to the group was to actually fly the mission on the computer."

This was crucial, they felt, in a couple of ways. First, it would be important to give NASA headquarters some confidence that the experiment they planned to do would have a fair chance of succeeding, roughly as engineers at a company like Boeing would build a new airliner on a computer to convince management that it was worth building for real. The other reason for building a virtual satellite was that the team could create maps of the sky in advance, based on different theoretical assumptions—a flat universe, a positively curved universe; varying densities of dark and ordinary matter; a cosmological constant, or not; and so on. When the real data came in, it would be easy to compare them with the simulated sky to see which mix of cosmic parameters best fit the actual universe—or, if some bizarre new physics happened to be going on, that none of them did.

The original team approved by both Bennett and Wilkinson had two other members, both of whom had worked with him on COBE: Steve Meyer and Ned Wright, the man who had served as that project's resident gadfly. "Ned," says Bennett, "is hard to categorize." Most physicists and astrophysicists are either theorists, like Spergel, or experimentalists, like Wilkinson, who build and use instruments to gather data. Wright and a small handful of others are a little of both. "He's right on the borderline," says Bennett. "He mostly does theory, and he's brilliant when it comes to software, but he knows a lot about hardware too. Not that he comes here and turns a screwdriver. But he's very hardware savvy." Equally important to both Bennett and Wilkinson was their common understanding of Wright's character. They knew that if you tell Ned Wright something can't be done, he'll stay up all night figuring out how to do it—a quality that would later prove crucial in convincing NASA that the project would really work.

This dream team wouldn't accomplish a thing, of course, without an approved mission. And at this point, there was no guarantee, first of all, that NASA would be willing to pour in the millions of dollars required, and second, that this would be the team selected to build it if they did. Both Bennett and Wilkinson knew that if and when the agency was willing to entertain proposals, they'd better be ready. It made sense to keep working out as many details as they could, just in case. But despite the small infusion of cash it had given them to begin studying the idea of a microwave background satellite, NASA wasn't encouraging scientists to dream up expensive new projects. The agency wanted to convince Congress it could control costs. If a spacecraft didn't fit into one of the agency's newly organized system of categories, it would have a tough time getting launched. By March 1994, though, the administrators at NASA headquarters were beginning to talk about a new category of missions that might include something like what the Bennett-Wilkinson team was proposing. The program would be called "MIDEX," for "mid-level explorer." (There was already a SMEX program, for "small explorer"; these inexpensive missions would include HESSI, for example, which would study solar flares, and GALEX, which would look at star formations in both nearby and distant galaxies. It was clear to Bennett that there wouldn't be any "large explorer" category.) According to the agency grapevine, the program would be looking for missions that could be launched in a year or two from approval, which appealed to Bennett. That was just the kind of mission a small, talented, focused team should be able to pull off—in contrast to COBE, which had taken eighteen years from initial conception to final results.

Even before the MIDEX program was formalized, Bennett began talking to headquarters, trying to get a sense of what it would look like, and offering his own suggestions about what might be most useful to scientists. Originally, for example, the people downtown were talking about using a Taurus rocket as the launch vehicle. But Bennett argued that the Taurus was too small. "You can't get a very big telescope in there for any kind of astrophysics," he says, "so it's hard to imagine how that would be very useful to astrophysics. I told them I thought they should consider something with a bigger fairing [that is, a bigger container for carrying payloads into space]." He also asked them how they envisioned MIDEX missions in terms of the care they expected to go into building them. In NASA jargon at the time, space probes were informally classified by letter. "Class A," explains Bennett, "is where the national reputation is at stake, so we're talking about something with the quality of the Hubble Space Telescope." (That's why the agency was so profoundly embarrassed when the Hubble first proved to have a misshapen mirror.) Class B missions were important, but not as obviously so to the bill-paying public; COBE, for example was a Class B. Class C, which included the SMX missions, was a notch lower still. "And Class D," says Bennett, "was basically:

'Go to Radio Shack and build it, and if it works, great.'" It was clear from his conversations with headquarters that MIDEX wasn't going to be an A—but it wasn't entirely certain it wouldn't turn into a D.

As NASA deliberated its new satellite program, Bennett and his team kept trying to define its satellite. That included coming up with a name—a catchy one, if possible. Bennett remembers sending out e-mails suggesting and soliciting dozens of ideas; on March 8, according to his notes, Bennett suggested that they call it the Microwave Anisotropy Probe. "Everyone said, 'Yea, verily,' and that was that. The name seemed perfect: it described what the satellite was, and the acronym MAP emphasized that the idea was not just to measure the whole sky; it was to make a map of the whole sky, which is not quite the same thing. The only minor regret I have," says Bennett, "is that nobody in the world knows what 'anisotropy' means or how to pronounce it." (For the record, it means "departure from perfect smoothness.") "I was wearing a MAP T-shirt in the public library one day, and the librarian came over and read it out phonetically. She said, 'I'm a real fan of words. What does that word mean?'"

CHAPTER 7

How to Design a Satellite

On March 9, the day after MAP was christened, Bennett heard from headquarters. He was right about the Taurus rocket, they said. They'd wanted a small launch vehicle to try and limit the total mass of MIDEX probes; they hadn't fully appreciated that this would also limit the physical size. They also told him that MIDEX would be a class C project, and that they'd soon be holding a workshop to talk with interested scientists about what kinds of MIDEX projects the community might be interested in doing. Bennett didn't even show up. "This is a competitive environment," he says, "and I couldn't understand why anybody would want to advertise what they were thinking." NASA headquarters also formed an in-house study group to try and figure out MIDEX. One conclusion: there was no magic way to make missions cheaper. "They pretty much came to the conclusions we already had," says Bennett. "You have small teams, working in a small number of locations, give the PI the money and the authority and let him do the best he can."

By May, headquarters had come around to Bennett's way of thinking about rockets: instead of a Taurus, the MIDEX program would specify a Delta Light launch vehicle, which could carry a probe that was twice as large. It was also in May that David Spergel joined the MAP team. Bennett felt that they needed a pure theorist, not just to figure out what the actual data meant, but also to help guide the satellite's design from the beginning—to help decide what specific kinds of measurements were needed to extract the most useful information from the CMB. He pushed for Spergel, whom he knew from the work on alternatives to inflation for the generation of cosmic structure, and also from the post-COBE meeting he'd organized at Princeton two years earlier. Wilkinson and Page, of course, knew him as an extraordinarily talented colleague: Spergel's formal scientific work and his informal conversations with others in the physics and astrophysics departments had earned him a glowing reputation on campus.

Spergel himself suspects another reason he was tapped for the MAP team. "When the COBE result came out," he says, "and I said publicly that our theory of cosmic textures was dead, I think people were impressed." Scientists sometimes cling to their pet models long after they should have moved on, and he suspects that the MAP folks saw him as someone who would let the data guide his interpretation rather than the other way around. (Bennett, who first proposed that Spergel join the team, says that's precisely correct.) Spergel didn't have a clue that he was being considered, though. He didn't even know that his colleagues, Wilkinson and Page, were talking about a satellite. "Lyman said he wanted to talk to me," Spergel says, "so we went up to town for lunch. I remember distinctly that we were walking past Hoyt [a biochemistry lab] when he asked if I'd be interested. I was. I was also surprised and very excited, since I'd been wanting to focus on the CMB."

His first formal encounter with the rest of the team happened at Goddard a few weeks later, where the scientists and a couple of Goddard engineers tried to sharpen their idea of what a mission should look like. At just about that time, they'd learned that the European Space Agency had approved a CMB-mapping mission of their own, tentatively scheduled for launch in 2001 (it's now slipped to 2007 at least). "Their stated aim," says Bennett, "was to 'kill the problem dead,' to be so detailed and complete in these measurements that there would just be nothing left to measure." The ground-based experiments they knew were planned or already operating (and which some MAP team members, including Lyman Page and Norm Jarosik, were working on themselves) would be nibbling away at the problem of small-scale fluctuations. The Europeans were now going into space to finish the job COBE started. "So there was kind of this whole swirling mess of things going on at every scale," says Bennett.

That didn't worry him a lot, though, nor did it worry the rest of the team. They'd never planned to kill the problem dead. "It was a very deliberate decision on our part," said Wilkinson, "to do a design that wasn't going to set the world on fire, that was going to get up there, make some measurements, do a better job than we could do from balloons and the ground. Doing something more ambitious would be a useful thing to do someday, but that would require a very complex, very expensive satellite that would take a long time to build." MAP, they figured, would fit into an unoccupied niche in both time—after some of the Earthbound experiments but before the European space mission—and scientific usefulness—far more comprehensive and accurate than the former, not quite as detailed as the latter.

In order to squeeze into that niche, the team had to be very hard nosed and specific about exactly what MAP's intended mission would be. A few weeks after Spergel joined the team, on June 20, 1994, the group gathered at Goddard—the science team plus Dave Skillman, who had fielded Wilkinson's questions so handily the previous fall. There had, by now, been months of e-mails and telephone conversations and teleconferences. Now they would spend an entire day going carefully through what the mission would be and what it wouldn't. "We sat down," says Bennett, "and asked ourselves, what are the 'need-to-do' things for this satellite, and what are the 'want-to-dos.'" Even if MAP wasn't going to kill the problem dead, they'd have to make trade-offs, giving up the ideal level of performance because the hardware it would require cost too much or weighed too much or wasn't reliable enough.

One key question they had to answer early on, because it would dictate much of the satellite's design, was what sort of electronic detectors would lie at the heart of its radiometers. One choice was to use bolometers, devices that warm up by a precise and measurable amount when they're hit with a microwave signal. The other was to rely on a kind of electronic amplifier known as HEMTs, for "high electron mobility transistor." Page, who had worked with both types of devices, favored the latter. "HEMTs are so much simpler," he says. "You can hit 'em and they'll work; you can drop 'em and they'll work." More important, they don't need to be cooled to within a few degrees of absolute zero, so the satellite wouldn't have to carry a bulky cooling system and liquid helium—a ferociously tricky substance to work with—to make it run. Bolometers do. Just to make sure they were being open minded, Bennett had Steve Meyer, who'd worked a lot with bolometers, argue the case for a bolometer mission, while Norm Jarosik and Lyman Page presented the case for HEMTs.

Bolometers did have one important advantage, though: they could be made to be more sensitive than HEMTs. But that was far outweighed by their disadvantages. The cooling problem was just one: not only would it make the satellite itself far more elaborate and failure prone, but it would also make the design and construction much more expensive. "You have to have a Dewar [essentially, a huge Thermos bottle] to hold the helium," says Jarosik, "and every time you transfer helium into it, you have to have an employee of the manufacturer there to observe. And since bolometers don't work at all at room temperature, all your testing gets very, very expensive"—a significant drawback in a MIDEX program where the budget cap was $70 million in 1994 dollars.

Beyond that, the highest possible precision wasn't the most important criterion for what the MAP team wanted to accomplish. What they wanted above all was for their final results to be as free from errors as possible. That meant that they had to be able to minimize sources of extraneous signals contaminating the pure cosmic radiation, and that they had to be able to account for and correct for whatever was left. "Often," says Bennett, "in ground and balloon-based experiments, people have to guess at their levels of error." That's because there are so many sources of error that it's difficult or impossible to measure and account for them all. On COBE, by contrast, Al Kogut had done an extensive analysis and written up a forty-page paper that said, in essence, "Here's exactly what our margin of error is, and here's how we know it." Other groups had been forced to issue retractions of or corrections to their original claims. The DMR team had never had to do that. "That's really what we want to do again, at that level of 'this is something I can take to the bank.'" ("MAP's errors paper will be a really, really important document," says Spergel, "even though very few people will take the time to read it.")

One powerful way to minimize errors is to make your measurements differential, just as the DMR had done. With bolometers, you couldn't do that easily. So in a bolometer mission, you'd have to measure the temperature in one direction, then swing the satellite around and measure the temperature in another direction, then compare the two. Each measurement would be very precise, but the comparison would be somewhat unreliable. "The problem," says Jarosik, "is that the response of your instrument can drift a little over time." Imagine that you want to compare the height of Mount Washington, in New Hampshire, with that of Mount Everest. You could take an altimeter to the top of Washington, make your measurement, then ship the instrument to Nepal and have a Sherpa carry it to the summit of Everest. But you'd run the risk that the altimeter had gone out of adjustment during the trip. During the interval when it's swinging over to another point in the sky, a bolometer's sensitivity can wander very slightly as well—and when you're trying to measure temperature differences a hundred thousand times smaller than the temperatures themselves, it doesn't take much drift to throw you off.

In a differential experiment, though, you don't have that problem. The instrument can't drift between measurements because no time elapses. "If you want the most possible sensitivity," says Bennett, "you would measure the total temperature at each point. If you want to believe your answer at the end, you do differential measurements, comparing every point to every other. And because we had made that trade-off already—sensitivity is nice, but having no significant errors is essential—we had no doubt of what we wanted to do."

It was also clear that they needed to measure the CMB at multiple frequencies—Hinshaw's simulations had shown that five was the optimum number—to make certain that any features that might appear in the microwave background really came from the microwave background. An effect that was more local, like emission from the Milky Way galaxy, should be much more pronounced at some frequencies than others. An effect that showed up equally strongly at five different frequencies, and from all directions, was more likely to come from the Big Bang.

The satellite also had to be sensitive to polarization—that is, to the orientation of the oscillating waves of electromagnetism that make up the CMB. The idea here would be to find evidence of reionization. Sometime after recombination and decoupling, the event that freed light from the charged nuclei and electrons that pervaded the entire universe until about 300,000 years after the Big Bang, the process was partially reversed. Electrons were stripped from hydrogen atoms again, by ultraviolet radiation streaming from the very first generation of stars: supermassive, fast burning, and very hot. Astronomers know reionization happened because light from the most distant quasars is partially obscured by clouds of ionized hydrogen gas. The question is when it happened: near a redshift of 6 or so, where the effect becomes pronounced? Or long before that?

So far, it's been impossible to say. But the signature of reionization will be embedded in the data. Think of reionized hydrogen as being analogous to air molecules in the Earth's atmosphere. When sunlight hits those molecules, it scatters (blue light scatters more than red; the daytime sky is blue in all directions because blue light is bouncing off air from all directions). But it's also polarized: the waves of electromagnetic energy that make up sunlight are bounced in such a way that they're oriented in one of just two directions, at right angles to each other. They are, to use the technical term, polarized. That's why Polaroid sunglasses work (and why they have that name): the lenses are inscribed with ultrafine lines, like a microscopic Venetian blind. They only let the light oscillating in one direction get through, cutting down the light entering your eyes by 50 percent.

Microwave photons would have been similarly polarized by hydrogen ions floating around in space after reionization took place. And their satellite, the team decided, should be designed to detect that polarization—not in great detail, because that would require elaborate instrumentation, but enough to measure how large, in angular size, the amounts of microwave sky polarized in one direction. If the patches were significantly bigger than two degrees or so, it would mean they couldn't have been generated at the time of recombination and decoupling. Polarization from that era could be coherent only across a span that light could have traveled since the beginning or about 300,000 years. Anything larger had to come later.

The third "must-have," they agreed, was that the satellite should map the entire microwave sky, in all directions, just as COBE had done. That was partly because nobody had ever completely verified the COBE DMR measurements, although enough ground-based observers had verified enough of them to make everyone pretty confident that they were right. Still, a second full-sky map that covered COBE's ground, from seven degrees on up, was valuable. But it was also important for another reason. There was a gap in angular scale from about one degree, the largest angle measured from the ground, to seven degrees, the smallest angle measured by COBE. If you didn't bridge that gap with a single experiment, you'd never be able to calibrate the one against the other. Again, imagine that you wanted to measure the heights of the Himalayas relative to the Appalachians. If you used two different altimeters, you'd never be certain they were calibrated exactly the same, so a thousand feet in altitude in the United States might not be the same as a thousand feet in Tibet. Using one altimeter to measure both mountain ranges would assure you of a valid comparison. "You've driven that one source of possible error to zero," says Bennett.

Finally, they agreed, they needed to produce a real, accurate map of the sky, not just make the measurements. "You can compute any statistic you want from the map," says Bennett, "but the map is the fundamental data." If you're showing signals from the sun or the sky, it's not accurate. If you're picking up signals from the instrument, it's not accurate. In telescopic photographs of stars, you often see spikes of light pointing up, down, and out to the sides. "The spike isn't in the star," says Bennett. "It's obviously from the apparatus. What the apparatus is doing is taking some of the signal from one place and putting it somewhere else. We didn't want to be doing that. We wanted the picture you see to be what's actually there." In an actual map, moreover, you would be able to pick up features in the sky that a mere statistical analysis would miss. Say the ripples in the microwave background were, against all expectation, twice as prominent in the northern sky as they were in the southern. A simple calculation of the overall, average height of the ripples would never reveal that astonishing fact. A map would show it at a glance.

Those were the goals, anyway. It would take a fair amount of work to figure out precisely how to accomplish them. For that, Gary Hinshaw's talent for doing simulations would prove essential: he could test out all sorts of virtual scenarios to explore such essential but mundane problems as how many detectors MAP should carry, how they should be arranged, how the satellite should spin, and even where it should orbit. COBE was designed to be launched aboard the Space Shuttle, so it had to stay pretty close to the Earth—in a polar orbit, specifically, swinging first over the Arctic, then the Antarctic. Anyone who's seen photographs of Earth from the Space Shuttle knows how large and bright the planet looms; COBE always had to point away from that giant source of contamination that eats up nearly half the sky. It also had to point away from the glare of the Sun, so the satellite was tightly restricted in the band of sky it could usefully observe. In a single orbit, it could scan maybe 20 percent of the sky.

The MAP team didn't want to be limited this way, so, says Hinshaw, they began scouting for other locations. "Dave Skillman," he says, "came up with a few possibilities." One idea was to put MAP into a highly elliptical orbit, a very long, very skinny loop that sent it repeatedly out by the moon and back. Most of the time, the Earth would be pretty small in its field of view. But even better, they figured, and Hinshaw's simulation confirmed, would be to send MAP out to a point in space known as L2. This is one of the five points described by any two mutually orbiting bodies, first identified by the mathematician Joseph-Louis Lagrange, who died in Paris in 1813. The salient fact about Lagrangian points is that they're islands of relative gravitational stability. A space probe placed in one of the Lagrangian points will tend to stay there, with minimal help from rocket thrusters. "It's kind of like going to India," says Spergel. "The price of getting there is high, but the cost once you're there is small." Of the five Earth-Sun Lagrangian points, three, labeled L3, L4, and

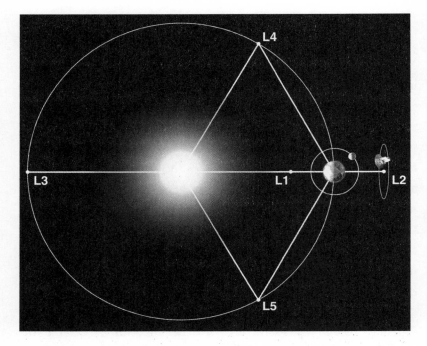

The Microwave Anisotropy Probe orbits at a point known as
L2, one of five Lagrangian points where objects tend to hover
stably with respect to the Sun and the Earth.

L5, are much too far away: one is on the opposite side of the
Sun and two are sixty degrees ahead of and sixty degrees behind
Earth's position in its orbit—tens of millions of miles away. L1,
which lies on the Earth-Sun line in between the two, is only
about a million miles away, so it's much more accessible. But
since the Earth and Sun flank it, one in each direction, avoiding
the glare of both forces you to limit your field of view.

That leaves L2, which lies on the same Earth-Sun line, but on
the portion that extends beyond Earth. Here, Earth and Sun are
both in the same direction; whereas COBE could see less than a
quarter of the sky in ninety minutes, MAP would be able to see

more than a third—a huge swath of sky that would slowly change as MAP, along with the Earth, made a one-year circuit around the Sun. Like COBE, MAP would spin, to take as many measurements within the available patch of the heavens as possible. Just how it should spin—how fast, and whether MAP needed to have more than one axis of rotation—wasn't so clear. Maybe it should tumble as well as twirl, to make sure the differential measurements would compare each patch of sky with every other in the most efficient way; that way you could get as many independent measurements as possible and heighten the precision of the final map. "I remember at one of those Goddard meetings," says Lyman Page, "that we were talking about various spin patterns. And some guy who was there—I forget who, but he's not on the MAP team now, just a smart guy in the audience—said, 'What if you do it this way?'" He sketched out a combination of spins that would trace out a sort of Spirograph pattern on the sky. "It clicked. I mean, he said just the right thing at just the right time." A run on Gary Hinshaw's MAP simulation software confirmed that this was about the best scan pattern you could put together. By now, David Spergel was working closely with Hinshaw, helping create virtual maps of the microwave background based purely on what theory predicted, so that the simulator would have something to "look" at.

As MAP inched forward from a concept to an actual design, meanwhile, NASA was inching toward an actual launch program it might fit into. On July 27, Bennett got word that the agency would issue an NRA (even Bennett, who should have all this memorized, has to think a moment before he remembers this stands for "NASA Research Announcement"). It was a sort of fishing expedition that would give the agency an idea of what kind of projects scientist were interested in doing. If it sounded promising, NASA would hand out a little research money, as Goddard had done earlier.

The people at headquarters were somewhat lukewarm when Bennett talked with them about entering his cosmic microwave satellite into the NRA competition; it was, they felt, not on a grand enough scale for what they had in mind. He decided to do so anyway, and ended up winning one of the dozen or so grants the agency ultimately gave out. When the NRA proposals came in, NASA administrators were chagrined to see that there wasn't a single proposal for a space telescope. "Rumor has it that they were concerned about that," says Bennett, "because they really wanted to plan for what would happen in the era after the Hubble Space Telescope. I think that what happened is that while the astrophysics community wanted it too, it seemed too obvious. What would you say about it in a proposal, anyway? So they were a little chagrined—they had to push what ended up as NGST [that is, the Next Generation Space Telescope] from their end. It was, like, 'Hey, community, don't you want a post-Hubble telescope?'" Headquarters snapped John Mather up to run that project, so, although Bennett hated to lose him, he had to give up his MAP affiliation.

Just to make matters more confusing, this NRA wasn't really aimed at the MIDEX program, which was developing along a parallel track. NASA had decided that the MIDEX would start out with two missions, and that they'd be selected in a two-stage process. First, any team that wanted to do a mission would submit a general proposal on the science it wanted to accomplish— what a satellite would measure or look at and why it was important. Whoever survived that cut would go on to round 2: showing in detail how you proposed to do it. Bennett didn't really approve. "Their idea was that you don't make a lot of people do all these details if you're not going to approve the idea to start with. My objection was that without knowing the details, you couldn't tell if an idea was doable, so it didn't matter how compelling it was." Nevertheless, a two-stage process it was going to be.

Bennett also had to find out how much support he was likely to get from his bosses at Goddard. As a staff scientist, he was encouraged to look around for interesting projects; that's why the Center had given him preliminary study money a few months earlier. But while he himself could participate in a non-Goddard project, he felt it was important to build it there, where the engineering staff had a lot of experience thanks to COBE and where, more important, he could oversee MAP on a day-to-day basis. "Unfortunately," he says, "just because we were sitting at a place where we were surrounded by thousands of engineers, that didn't mean they would automatically work on anything you wanted." He had to talk the Center administration into it first. So on August 10, he made a presentation—as did several other scientists with other projects—to try and sell MAP. Evidently, it worked. "They decided it should be given the absolute highest priority."

Proposals for the MIDEX program were due the following spring; ultimately, headquarters would get about sixty of them, for missions designed to explore everything from the Earth's magnetosphere to the origin of galaxies. There was a general sense in the air, says Bennett, that one of the two missions that was finally selected would almost certainly be a microwave-background probe, because the science was so obviously compelling. "I'm sure at some point the headquarters guys virtually told me that," he says, "but I'm also sure they didn't tell me verbatim, because that wouldn't have been allowed." It was no secret, however, that two other groups, one at JPL and another based at Caltech, were also proposing cosmic-background experiments. The former, called PSI (for "Primordial Structure Investigation") came from a team at the Jet Propulsion Laboratory and was based on actively cooled HEMT amplifiers; the latter, called FIRE ("I think it stood for Far Infrared Explorer," says Bennett) used bolometers. Neither mission was differential.

Calling the first round of reviews the "science" round is somewhat misleading. It was true that the teams didn't have to specify every last detail of their proposed missions, but they did have to demonstrate that they could deliver it on time—launch had to come by April 2001—and within the $70 million budget. To do that convincingly, they at least had to present an outline of the technical specifications, management plan, and cost estimates. In the end, FIRE, MAP, and PSI all made it past that first review but Bennett heard rumors a few months later that NASA's science advisory panel had wanted to kill FIRE at that point. "This definitely wasn't from headquarters," he says. "It percolated in a sort of some roundabout way through the community. To this day, I don't know if it's actually true." In any case, NASA kept them all alive, presumably because measuring the CMB was seen as so crucially important.

That first proposal had been tough enough to compile—thirty pages long, each one dense with information. The second, though, was immeasurably worse. It was due in December 1995 and, says Bennett, "I'll never forget that period as long as I live." "I asked my wife, 'Are you really willing to go through this? Because you might not realize how much of an impact it's going to have on our family life.'" In fact, he pretty much didn't see her or their two young sons for three months straight. "I would come in to the office early in the morning, and work on the proposal. We would work on graphics and wording and stuff all through the day. The office staff would put in their eight or ten hours and go home. I'd keep working. I'd print out a draft and then bring it home, at maybe one in the morning or something. And I'd read through it and mark it up until three. I'd get four hours sleep and then back in the next morning and putting those changes in and moving on."

It was almost as hard for the rest of the team—especially, he says, for Cliff Jackson, the Goddard employee who had signed on as MAP's systems engineer. Jackson, a slim, soft-spoken, and almost invariably smiling man in his late fifties, had been at Goddard since 1970, after graduating from Princeton. At the time MAP was gearing up, he was looking for a new project to work on, since his most recent, FUSE (a far-ultraviolet mission), had been "transitioned" out from under him, out of Goddard and over to the Applied Physics Lab at Johns Hopkins University. During his struggles with George Smoot, Chuck Bennett had learned that the best engineers, the ones most in demand at Goddard, were motivated not just by engineering challenges but by the science they were helping to do. Luckily, Cliff Jackson was fascinated by MAP. "I liked the cosmology," he says, "and I liked the fact that we were working with Princeton, where I'd gone. I sort of had a choice between a couple of projects. I don't remember what the other ones were now." He also wanted to work on a project that actually made it into space—something he hadn't done in nearly a quarter-century at the Center. He remembers that one seasoned Goddard manager told him of MAP: "Oh man, it's a little early for this one . . . I don't think it's really gonna make it. It's good but it's just not likely to fly." Fortunately, Jackson didn't agree.

Another reason he signed on with MAP was that Chuck Bennett was the PI. Jackson's daunting job would be to make sure that all of the systems in the spacecraft—electrical, mechanical, thermal, optical—worked individually, and worked together, and let the actual microwave-detection instrument do its job with a minimum of interference (and, moreover, make sure that any interference was precisely understood). He would have to juggle the interests of the scientists, the engineers, and the project manager, whose job was to keep the whole thing on schedule. "Some PIs," he says, "are focused entirely on the science and forget about the other, and you get into loggerheads with the project manager. Chuck really takes a global view; you know, to be successful the whole thing has to work. Scientifically, programmatically, it's all got to work. Chuck's practical," he says— the ultimate compliment an engineer can pay a scientist. Jackson and Bennett worked side by side through both the Phase 1 and Phase 2 proposals. "Cliff was exacting and tireless," says Bennett, "with only occasional breaks to go outside and kick a soccer ball against the building."

A huge part of the process involved checking in with the people at Princeton, figuring out what the weak points in the proposal were, the holes that needed to be filled, the answers to questions the reviewers might ask. Why are we saying this? What's the argument for that? "Cliff or I would be on the phone constantly," says Bennett, "with Lyman and Norm and Dave Wilkinson." Lyman Page was appalled at first, especially at Cliff Jackson's persistence: "What I've learned about systems engineers," says Page, "is that at first they just drive you up a wall with one seemingly insignificant question after another. And then after a little while you realize they're God. And everything they've been saying is right—all of those questions are incredibly important; they're not just designed to drive you insane. I talked to Cliff Jackson more than I've ever talked to any other human being in my life. And the amazing thing is, I know he talked to other people just as much as he talked to me."

Page's job was dealing with the satellite's optics. He was the one who had to figure out how to take the microwaves streaming in from space, bounce them into twenty different antennas, two for each differential radiometer, at five different wavelengths. He had to figure out what shape the radiometers would take. They'd be horn shaped, of course—something like the bells of brass instruments, only elongated—for this had long since been shown to be the most effective way of funneling microwaves into an instrument. But precisely what they would be made of—some sort of carbon composite with a metallic coating, to save weight and maintain a perfect shape as outside temperatures changed? Aluminum for strength?—wasn't clear. Neither was how they'd be arrayed on the probe. In essence, MAP was a microwave telescope, and there's no way to put all the horn antennas at the optimal focal point. So the aluminum-coated composite mirrors, a primary and a secondary, had to be shaped to give a pretty good focus to all the horns, but not a perfect focus to any of them.

Page also worked on MAP's thermal design, but most of that effort happened at Goddard. "Early on," says Jackson, "we thought, 'Hey, this is gonna be easy.'" The front end of the satellite had to be kept as cold as possible; while radiometers worked at any temperature, they worked most reliably and accurately when they were cold, and the colder the better. In deep space, you could get things pretty cold simply by shielding them from the Sun. But the satellite would be drawing power from solar panels that faced the Sun, and the power had to get from the hot side to the cold side, so the solar shield had to have holes in it, which had to be carefully designed. The cables themselves would get warm no matter what, though, so some solar heat would leak through. Not a terrible problem as long as the heating was steady and didn't trick the radiometers into thinking they were seeing a cosmic variation. But it wouldn't be perfectly steady, because the solar shield and the solar panels couldn't be perfectly symmetrical, and their surfaces couldn't be perfectly even in how much heat they absorbed. As MAP rotated, there would be subtle changes in how much they heated up, depending on which parts were getting the most direct sunlight.

The spacecraft had to carry heaters, moreover, because the fuel that fed the thrusters it used to stay at L2 could freeze; the radiometers had to be shielded from these sources of false detection as well. They also had to be shielded from the tiny surges of heat that would be produced by reaction wheels. By changing the rate of their own rotation, these gyroscope-like wheels would be used to set and maintain the spin of the spacecraft—but the changes would draw spikes of power.

All of these had to be shielded against and, if the shielding was imperfect (as it would inevitably be), accounted for. If the mission was selected, every conceivable source of thermal noise would be rooted out in exhaustive prelaunch testing. But since things might be different once MAP had been shaken around during the launch and cooled to deep-space temperatures, it would also carry thermometers—thirty of them, designed by Steve Meyer and located at strategic points around the probe to measure changes as small as half of a thousandth of a degree.

The most exacting job of all, and the one that would take longest, was the construction of the radiometers themselves. They were hardly something the MAP team could order out of a catalog, given the extraordinary sensitivity and robustness and reliability needed to make millionth-of-a-degree measurements in deep space. They'd have to be handmade, starting with the eighty HEMT amplifiers at the heart of the twenty radiometers. The radiometers would be assembled at Princeton, but the amplifiers would clearly have to be contracted out, since neither Norm Jarosik nor anyone at Goddard had the expertise for this highly specialized and exacting job. One option was to go to one of the aerospace companies that routinely build electronics for spacecraft, but in the end Bennett and Wilkinson decided to approach the National Radio Astronomy Observatory, whose headquarters is in Charlottesville, Virginia, at the foot of the Blue Ridge Mountains, instead. NRAO's entire existence was dedicated to making long-wavelength observations of the universe, from microwaves to radio waves; among the observing facilities it runs, wholly or in part, are the Very Large Array in New Mexico; the Arecibo Observatory in Puerto Rico (both featured in *Contact* with Jodie Foster); and the Green Bank Observatory in West Virginia.

NRAO has its own electronics shop, with a local genius, Marion Pospiezalski, a legend among radio astronomers for his skill at designing components that generate as little noise as possible. In this case, he'd have to create amplifiers that would pick up the portion of the CMB spectrum at 90 gigahertz, a region of the spectrum where intereference from the galaxy would be especially low. That, says Bennett, had never been done—another reason to go with a Michelangelo of electronics like Pospiezalski. (His faith wasn't absolute, though: using some of the concept money Goddard had given them initially, Bennett and Wilkinson went to Hughes Aircraft and bought a wafer of HEMT transistors, handed them over to Pospiezalski, and asked him to make a prototype of the high-frequency amplifier. When the proposal said it could be done, they wanted to be sure it really could. Evidently, NASA headquarters called during the second-round selection process to ask Pospiezalski what he thought. "I think it's doable," he told them.)

In November, as the deadline for the final proposal loomed, an event took place that would give the MAP team another boost in its confidence that it could actually do what it was proposing. As they'd decided a year and a half earlier, MAP would be a differential experiment, making a direct comparison between every possible pair of points in the entire sky. The number of individual calculations involved was huge, even for COBE, where the "points" were seven degrees in diameter, or fourteen times the width of the full Moon, for a total of about 6000 patches of heaven. MAP was proposing to look at fifth-of-a-degree patches, for a total of 3 million. Comparing all of these to all of the others was a computational nightmare. In the meetings where they'd decided on many of MAP's basic attributes, the team had simply waved its hands on this issue. They had no idea how they'd do the calculations. They basically just said, "Computers will get faster, and we'll figure out something."

Maybe so, but most astrophysicists familiar with the problem were sticking with the proposition that it was pretty much impossible. That's what Ned Wright kept hearing at a cosmology conference he attended at the University of California, Santa Barbara. Sitting in the audience, he kept thinking about these claims that you couldn't analyze a small-angular-scale differential sky map. It bugged him: as Chuck Bennett had once said, Ned Wright doesn't take well to being told something can't be done. So he began to formulate a way to do it. He was savvy enough not to talk about his idea with anyone else at the confer-

With a resolution of 7 degrees of arc, COBE can pick out only
the grossest features of the cosmic background radiation (image
at top). MAP's 0.2-degree beam will uncover far subtler patterns of high and low density (simulated image at bottom).

ence; no point in tipping your hand to possible competitors. But
on the drive home on the final day of the conference, the idea
jelled. "It's about a hundred miles from my house," he says,
"and it's anticommuting. So you get on the freeway and drive
past all the people stuck in traffic jams. Takes about an hour
and three quarters." By the time he'd reached home, Wright had

sketched out an algorithmic shortcut that he was pretty sure would dramatically cut the processing time without sacrificing accuracy. "I wrote an e-mail to Chuck and Gary Hinshaw and the others right away explaining why it was a good idea and why it would make the process fit a lot better on existing work-stations."

When Hinshaw saw the e-mail, he remembers, "It was, you know, 'bingo,' as if a white light had gone off and said, 'Yes, this is how to do this.'" Hinshaw was very comfortable with his mission simulator by this time, and it was very simple to adapt it to run Wright's algorithm. "I saw the e-mail at about nine in the morning," he says. "I had the code modified by noon, and I had the first few iterations of the output ready by about six o'clock that evening. I sent out an e-mail to the team saying how well this worked." The technique actually made its public debut in the Phase 1 proposal, although Wright formally wrote it up for a scientific journal soon thereafter.

In the end, the Phase 2 proposal would be two inches thick, and just as dense with specifications, costs, and other details as the earlier document. "It was a really, really serious proposal, unlike anything I had ever done before." But Bennett knew that NASA would be scrutinizing it with extraordinary care. In its desperation to survive in the face of budget cuts and hardware failures, the agency was basically asking proposers to design missions that were as inexpensive as possible yet posed a minimal risk of failing to do what they promised. "I figured that one of the ways that they would judge risk was by how much you'd thought about it. The more detailed the proposal was, the more confidence they would have that we had thought about it, and knew in great detail what we wanted to do."

The proposal went to headquarters in December 1995, during a government shutdown triggered by one of Newt Gingrich's head-to-head confrontations with Bill Clinton. "Everyone was officially on furlough, so we weren't supposed to be in the office," recalls Cliff Jackson. "It was a real circus, trying to get the thing printed. And then Chuck and I drove it to headquarters, which was deserted. We had to carry the boxes in ourselves, try to find the right place to take them. Because shutdown or no, they weren't extending our deadline by more than a day or so. It was . . . interesting."

Officially, there wasn't anything to do while they waited for NASA's review board to pass on their proposal. Unofficially, they acted as though the mission had already been accepted—not, says Bennett, from overconfidence, but because it would put them that much farther ahead if it did happen. Not only did they have Marion Pospiezalski start figuring out the amplifiers: they also put together 150 web pages on cosmology, the foundation for an eventual public site that would tell the public all about the mission and what it was trying to do. "We would have to do that eventually," he says, "and we didn't want to be dealing with it when we had actual hardware deadlines dangling over our heads."

Even though he kept himself busy, Bennett found the wait agonizing. "There were all kinds of rumors floating around the community—this one's going to win, that one's going to win—I heard this, I heard that. It was all pretty much a bunch of nonsense." It was clear, though, that one of the three CMB missions in serious contention—MAP, PSI, or FIRE—would be selected. Headquarters would say they planned to decide by a certain date, and the date would come and go, and they said nothing.

"Then there came a point," says Bennett, "where they said they'd made a decision but they couldn't tell anyone yet. They had to notify Congress first and get formal permission. It was sort of like . . . torture." But one day in mid-April 1996, he finally got a call from headquarters. MAP was formally approved. "At the time," he says. "Wes Huntress was head of space science, and the way I hear it, he called Joseph Rothenberg, who was Goddard's director at the time, and told him the selection was a no-brainer and that the mission was worth twice the price. To which I understand the answer was: 'I'll take it!'" The advisory committee that chose MAP liked the fact that, aside from its 90-gigahertz amplifiers, it used no new technology. FIRE, on the other hand, was a bolometer-based satellite, and PSI was downgraded for having to use active cooling. "It was much higher in cost," says Ray Weiss, who sat on the committee, "and we thought that its reliance on cryogenics to cool the detectors was too risky. We also liked the fact that MAP was a differential mission."

There was one more hurdle before Bennett could consider the selection absolute. MAP had to pass what was called a Confirmation Review, a bit more than a year later. The MAP team had to produce what was called a "definition study," which would nail down the technical details, and especially the costs, much more specifically than they'd done to date. It wasn't unheard of for a mission to be selected, then fail this review and be unceremoniously canceled. In fact, the MIDEX program administrators had selected a backup mission to replace it if this should happen. "It was an ultraviolet experiment," says Bennett. "The PI was Richard Henry, from Johns Hopkins. He was very gentlemanly—called to congratulate me, told me he assumed we'd pass, and that he wouldn't rock the boat in the meantime."

According to the MIDEX program rules, MAP had to launch no later than March 2001. But Orlando Figueroa, head of the Explorer program office, argued that the team should be told the deadline was August 2000. Bennett remembers saying "There's no way we're going to make August. And Orlando said, 'Yeah but let's aim for that anyway, because it'll make us manage more aggressively and force us to make decisions.' He had a point, but I argued that there was just no way, it wasn't even reasonable. You couldn't write down a schedule that would make that happen. And so we ended up—this was a friendly discussion, I should emphasize—we ended up with a compromise, that we would plan for a launch on November 7, 2000. We just split the difference between March and August."

Because of the tight turnaround, Bennett and Rich Day, the project manager, knew that they couldn't concentrate solely on the definition study. If they were going to make the launch deadline, they'd have to get moving. They had to put together an engineering team of one hundred people or more and figure out who was doing what. And they had to formalize contracts with anyone outside Goddard who was working on the satellite—a process that he knew from experience would take a lot of time, and which could blow the entire schedule if it were allowed to lag. Some of the outsiders were specialized contractors like Lockheed-Martin, which would be building the "star trackers" that kept the satellite properly aimed by triangulating from known stars. But NRAO and Princeton were also considered contractors: they had to go through this highly ritualized drill as well. Or something like it, anyway. When Dave Wilkinson received Goddard's standard contract, the deeply antibureaucratic physicist reacted predictably. "That thing was five inches thick," he said, still appalled at the memory of it. "I called Chuck and said, 'This is ridiculous. Send me a five-page contract.'" Bennett, with his insider's knowledge and post-COBE clout, was able to do so.

Bennett also had to write a quality-control protocol that would meet Goddard's rules about quality assurance. Those who build flight hardware for a space mission have to have their own quality-assurance team that guarantees that the contractor meets Goddard's standards for laboratory design, materials, assembly techniques, and a list of seemingly petty regulations that appear to have no other function than to drive people nuts. "Despite its bad reputation," says Bennett, "quality assurance is a good thing. The reputation comes from making people do things that are silly." Bennett vowed that this wouldn't happen with MAP, and, luckily, the quality assurance engineer assigned to the project understood. "He was," says Bennett, "a very sensible, insightful person. We relied on him heavily to separate the wheat from the chaff—to know what we could drop as useless, and what we had to do to avoid making dumb mistakes and failing performance reviews later on."

So they wrote up a plan tailored specifically for Princeton. Dave Wilkinson and Norm Jarosik found a couple of rooms in Jadwin Hall where they wanted to build the radiometers; the Goddard quality assurance engineer, Mike Delmont, came up and checked it out, listened to how Princeton would handle air filtration, cleanliness, and humidity. "I said, 'They know what they're doing.'" recalls Bennett. "'You can suggest techniques for dealing with these issues, but don't push.'" In some cases, Delmont's QA team gave Princeton some very useful ideas— for antistatic gowns, for example, which would keep potentially electronics-frying sparks to an absolute minimum. They also found a way to streamline the normally cumbersome record-keeping requirements so the Princeton folks could document their construction techniques adequately, but not ad nauseam.

The same sort of flexibility applied to NRAO. The QA team needed to know what parts they were going to use, for example, because NASA has a list of advisories on parts that are considered unreliable for space. "But we weren't going to tell them how to build this stuff," says Bennett. "These guys were the experts. But maybe there were things they were not so used to." Don't do things a certain way just because NASA has always done them that way, Bennett said. But don't throw procedures out without understanding what they're meant to accomplish. "Mike Delmont came to me any number of times," he says, "to assure me that they weren't coming up with makework. Here's what we're worried about, and why."

CHAPTER 8

The Build

Nobody involved with MAP would have much rest for the next five years: pretty much everything on the probe had to be designed, built, tested, and assembled on at least a dozen parallel tracks at once. Even so, the most agonizing crunches came in waves, hitting different members of the team at different times. One of the first waves would break at NRAO, where Marion Pospiezalski had to crank out 120 low-noise amplifiers—eighty for the spacecraft, and forty extras—as fast as he could. "It was a little painful for them, I think," says Bennett. "They really have more of a small research environment there, and they had to change it into a production environment."

Fortunately, Pospiezalski had an able deputy named Ed Wollack who was developing something of a reputation as an instrument builder himself. Wollack had come to NRAO directly from Princeton, where he'd worked with Wilkinson, Jarosik, and Page, among others, putting together ground- and balloon-based microwave background experiments. He'd originally come to Princeton hoping to work on a phenomenon in condensed-matter physics known as the fractional quantum Hall efffect, but the funding dried up by the time he got there. Dave Wilkinson offered him the CMB as a consolation prize. "I actually knew a little about it already," he says, "because I had spent a year at the South Pole between undergraduate and graduate school." Like Page's postgraduation sojourn, his job was to watch over other people's experiments, especially during the six months of pure, frigid, storm-ravaged darkness that dominates the Antarctic winter. "I was the fix-it guy," he says. "If anything broke, I figured out how to make it run—everything from optical telescopes to particle cameras. It was a great introduction to fieldwork: it's so cold and dark that there's a real premium on thinking things through carefully before you actually do the repair." One of the experiments that was being installed as he departed after a year on the ice was a CMB detector, built by Phil Lubin; it seemed pretty interesting, so he read up a bit on the field.

When Wilkinson broached the topic, Wollack was immediately intrigued. "They were very interested in making very precise tabletop kind of experiments of low-noise systems. And I looked at it as an excellent way to keep learning how to be an experimental physicist." He found Wilkinson to be a very exacting but gifted instructor. One of the first problems Wilkinson gave him was to go make part of a waveguide, a piece of metal of a precise shape that would channel microwaves from the antenna into the detector without distorting them. "I looked at the problem, and I was like, okay well . . . I designed it and figured out all the nuts and bolts. How to go get it made. Talked to people in the machine shop, and they were very helpful in telling me how to make it. And I came to Dave and I said, 'Well, here's what I want to do.' And he said, 'OK, go do it.' And when Wollack had finished, Wilkinson told him to characterize it—that is, to show on paper, based on the design, exactly how it would perform. And suddenly, Wollack realized that, after spending a month of his life on this task, the part would never work properly. Wilkinson had known all the time. "Dave didn't look at me and say, 'You screwed up,' or anything. He said, 'Now do you know what questions you have to ask yourself?' It was a really important lesson, because if you have to take a month, or a year, or three years in this business to make something, you do it. But you better know that when you get done you can do the measurement you want to set out to do." The other crucial thing he learned from Dave Wilkinson: "If I come up with an experiment that costs more money than the gross national product, they're never going to build it."

Armed with these real-world skills and the experience of getting several real-world experiments up and running, Wollack moved to NRAO for a postdoctoral fellowship studying the noise properties of HEMT amplifiers. Rigorous as his training at Princeton had been, Wollack quickly realized how little he knew. "Marion works with an extremely skilled technician named Bill Lakatosh. He comes up with the design and then he talks it through with Bill. So maybe Bill says, 'I can't make that,' and Marion says, 'OK, can you make this instead?' They go back and forth, and deal and trade and end up with a working design. I just watched them and I learned a huge amount." It took more than a year, in fact, before he could begin to appreciate Pospiezalski's genius. "I mean, he could look at something and see in three dimensions how it was all going to fit. Sometimes I didn't fully appreciate how brilliant he was until I'd finished putting something together."

Wollack had been at NRAO for a year and a half when the MAP order arrived for 120 amplifiers, including forty in the so-called "W" band at 90 gigahertz. "At that point," says Wollack, "Bill Lakatosh was only one person in the entire world who had made one, and as far as I knew there were only two people who had the skills to make them." Goddard wanted them fast, but it wasn't like NRAO could churn them out on an assembly line. It took a while for the technicians to get up to speed. "One of the guys worked for three years before he could put together a flight-ready amplifier," he says. "And it wasn't like he wasn't working hard at it." By the time they'd delivered the last device to Princeton, says Wollack, the number of competent technicians had increased from two to five.

Along the way, he also learned plenty about how to improvise. Early on, for example, he realized that the amplifiers would need to be equipped with tiny probes that received the signals from the antennas and channeled them into the HEMTs. "They're something like the whip antenna you use for a car radio," says Wollack. These bits of metal had to be shaped sort of like minuscule doorknobs with very long shafts only twice as thick as a human hair. NRAO had one machinist, named Matt Dillon, who could turn them on a tiny little lathe—but he couldn't make them fast enough to meet the schedule. "I'm trying to figure out how we can do it," says Wollack, "and Matt says, 'Well, we can get Françoise down here to work on that.' And I'm thinking, well, yeah, that's great, but I need her cutting these *other* pieces I can't get anyone else to make." He called up some electronics companies to see if they could do it—and they could, as long as he was willing to order a minimum 10,000 of them.

Finally, the secretary, who'd been watching Wollack go back and forth, asked how Matt made these things. He told her about the tiny lathe, and she said, "Well, why don't you call the company that makes the lathe and see if they can help?" He thought, "Yeah! That seems like a good idea!" They sent him to another company, which had customized the lathe, and Wollack asked for the name of some other customers—people who presumably made parts something like what he needed. They wouldn't tell, but they were willing to inform those other customers that someone in Virginia was interested in having them do some work for him. About two weeks later he got a call from a small company in California that made pieces for heart valves. He faxed out the specifications, and a couple of days later, five arrived in the mail. They weren't quite right, but Matt Dillon was impressed with how fast they'd been turned out. He got on the phone and talked the process through with the guy, and the next batch were perfect. "The cheapest quote I'd gotten from any machine shop before this was $250. He charged us something like $7.50 apiece."

Wollack also learned that at some point you have to be done, and you have to know when you're at that point. "There are three hundred bonds on a W band amplifier," he says. "And you go in under a microscope and you look at it and you go, 'Well I don't like that, that, that, that'. . . so you go and fix it. Then you increase the magnification and look again, and you find more problems. But at some point you've got to say to yourself: 'Is it gonna work? Or am I gonna make it less reliable by pulling it apart?' And that's something you can't teach people." In fact, it seems like magic at first. When he started, Bill Lakatosh might look at a bond and know it was wrong. "He could see that it was just a little less shiny-looking that it should be. To me, they all looked the same. But today I can look and they don't look the same."

Perhaps the most important lesson the MAP experience taught him, though, was how to get people to do what you need them to do. One example: all the technicians working on the amplifiers had to wear a wristband that drew off static electricity. If you weren't wearing it and threw off even a tiny spark, you could fry the whole thing. But one technician told Wollack: "I've been building these things for years, and this is a bunch of crap. I'm not wearing it." So Wollack consulted his most trusted advisor—his wife, Allison Pihl. Her advice: let the man build the "engineering units" that would be used for testing but not on the spacecraft. He fried a couple of them. After that, he wore the strap. In another case, a technician was putting layers of material on in a way that Wollack was sure would let them come apart—but there was no way to prove it except to put them through some testing. The layers curled up. The technician took Wollack aside and showed him, and asked, "How am I going to deal with this?" Rather than tell him, Wollack asked, "Can you think of some other way of attaching them?" The man came up with his own solution—not what the others were doing, exactly, but something that worked equally well. By the time the last amplifier went out the door, Ed Wollack had, to his own surprise and to his pride, turned into a reasonable facsimile of Dave Wilkinson.

As the amplifiers were completed, Wollack and Pospiezalski shipped them up to Goddard, where engineers would subject them to vibration testing, send them back to NRAO for performance testing and then, if everything worked, they'd go on to Princeton where Norm Jarosik and a postdoc named Michele Limon stood by ready to assemble them into twenty radiometers—one each for the lower-frequency channels, two for the higher frequencies, where the response is less sensitiver. "We were getting Heathkits up here," says Jarosik. "All we had to do was put 'em together."

Actually, it was just a bit more complicated than that. The amplifiers worked fine on their own; now they'd have to work fine hooked up together. The same sorts of danger—burnouts due to excessive voltage, mechanical failure of the assembly— were possible at this level too, except that the deadline was that much closer. Jarosik and Limon had to be careful to design the radiometers so they'd be as compact as possible, but also so that the assembly process was completely logical. There would necessarily be points in the process where some subassemblies were covered with others and couldn't be gotten to; if you covered something up and discovered later that you needed to attach one more wire to it, that would be a serious problem.

Given the tightness of the schedule that Chuck Bennett and Rich Day had set up, in fact, just about anything that slowed the assembly process could be a significant problem. Each radiometer took about three months to build, and Jarosik had three of them going in parallel at any one time. (He upped it to four at one point. The radiometers had to be cold-tested in a special chamber called a "toadstool" for its top-heavy shape, in order to compare their performance at room temperature with their performance at space temperature and thus predict how they'd behave on the actual mission. So when a shipment of amplifiers was late, he used the idle time to have the physics department machine shop design and build a fourth toadstool.)

But inevitably, things went wrong, just as they had for Ed Wollack. Probably the worst was the screw incident. The radiometers were put together with a special kind of screw, and NASA insisted that they had to be made out of a special alloy, to strict specifications. So Jarosik put in an order to a manufacturer who was experienced at making space hardware. They came in, and they looked fine, and worked fine—until one day, when Jarosik was tightening one, using a torque wrench to make sure he was applying precisely the right pressure. Suddenly, something was clearly wrong: instead of getting tighter and tighter, the screw started to feel "mushy," as Jarosik describes it. Then the head broke off completely.

This was very, very bad: he'd already used screws from the same batch on radiometers that were now finished. Maybe it was just one defective screw. He built a test rig, and tried a few more. Two snapped off. "At that point," he says, "we just stopped everything and called Goddard in. It turned out that the manufacturer had actually built the screws to the wrong specifications, and out of the wrong alloy to boot." It was possible, everyone realized, that the completed radiometers could fall apart—maybe under vibration tests at Goddard, maybe not until they went into space. "You'd think, OK, we'll just replace the screws," says Jarosik. "But unfortunately, once they're in, we coat 'em with a kind of silicone rubber, and then epoxy, to keep them from vibrating loose. If we took them out again, we'd have this gook in there, which crumbles when it dries, and goes everywhere. It would be impossible to get the joints mated perfectly again with that stuff in between."

After looking hard at the problem, the Goddard quality-assurance people decided it was less risky to leave the radiometers alone than it would be to try and fix them. Each connection used four screws. One could fail without fatally weakening the connection. Even two could fail, as long as they weren't adjacent to each other, and the chances of that happening, everyone figured, was pretty small. "I actually wasn't too worried," says Jarosik. "That mushy feeling I got from the one screw that did fail was pretty obvious. If the others had been defective, I think I would have felt it."

By now, Wilkinson had backed off from active work on the satellite and become more of a troubleshooter, helping Jarosik and Limon deal with screws and other nagging problems. He'd been in the microwave-background-detection game for more than thirty-five years now; it was time to let others start setting the stage for the next thirty-five. And he'd been diagnosed with lymphoma, a cancer of the lymphatic system. The chemotherapy was working, but it left him tired and weakened his immune system so that colds mutated all too easily into bronchial infections. (His idea of slowing down, though, might not correspond to most peoples': inspired by a passing comment a guest lecturer made in one of his courses, Wilkinson would soon launch an entirely new project. Along with Norm Jarosik and a gaggle of amateur astronomers, he would refurbish an unused telescope on campus, patch up the observatory around it, and begin searching for flashes of light that could be laser signals coming from alien civilizations—frequently pulling all-nighters to help train his volunteer observers. It was possible, though not likely, that Dave Wilkinson would discover alien life before MAP reported in.)

As Jarosik and Limon completed the radiometers, they'd send them down to Goddard for more shaking; if one of these things was going to break, everyone wanted it to happen on the ground, not in space. Then the Goddard engineers would put it through more testing. One day Jarosik got a call from Goddard: someone had failed to bolt a radiometer properly into its test rig. The device shook loose, but even though they stopped the test instantly, it was too late. "It was pretty well mangled," says Jarosik. "Somewhere or other we have pictures. Anyway, they were really worried and wanted me to come down." Instead, he had them ship the debris up to Princeton. Fortunately, the electronics inside hadn't been damaged, and he managed to put it back together without losing more than a couple of weeks. "Goddard helped us out of some jams a few times," says Jarosik, "so it was nice to be able to help them out of one."

While Jarosik and Limon were building radiometers, meanwhile, Lyman Page was working out the optics. Just as with an optical telescope, MAP would have primary mirrors that gathered the faint microwaves of the CMB and sent them bouncing off secondary mirrors into the horn-shaped antennas that would funnel the radiation into Jarosik's radiometers. By now he'd determined that the mirrors would be made from a carbon composite, with a thin coating of aluminum, for reflectivity. The antennas would be aluminum. Their shapes were pretty well fixed by the frequencies they were trying to detect, in the latter case, and the need to get the microwaves focused into the horns as effectively as possible, in the former. As a veteran by now of several ground- and balloon-based CMB experiments, Page knew how to deal with all of this.

What made MAP different was that in order to be as error free as possible, the science team had to understand every source of extraneous noise. For the radiometers, that could come from heat or electronic noise; for the optics, the main worry was external microwave emissions from the Sun or the Earth. "Even though the Sun is always behind you in this special orbit," he says, "it's really hot. Six thousand degrees. And we need to reject that radiation to just one part in a billion." To make sure they were doing it, Page had a mockup of MAP—all of the optics, but not the radiometers or electronics—built on the roof of Jadwin Hall. From there, it's easy to see, across the road and just slightly uphill, the rooftop where Dave Wilkinson and Peter Roll had built their primitive radiometer, and where they would have discovered the CMB if not for Wilson and Penzias at Bell Labs.

Page also had a bright light installed at the top of the math tower, an eleven-story mini-skyscraper that looms over Jadwin, and aimed it down at his apparatus. The light would simulate the Sun, and by moving the mock MAP around he could construct a kind of topographic map of precisely how much light spilled over into the satellite's vision, and from what angles, depending on how MAP was oriented. By mapping the satellite's so-called side lobes—its response to signals at the very edge of its beam, or field of view—he could give Gary Hinshaw, Dave Spergel, and Ned Wright a sense of how well they could trust the data they were seeing, and how much to throw out. By the time he was done, they'd understand the shape of the beam to within half a percent. Of course, the antennas, like the electronics, would be shaken up during launch, and they'd be heated and cooled many times, on Earth and in orbit, before settling into the frigid cold at L2, so they could conceivably change shape subtly. But that wasn't a problem, because several times during each data run, the satellite's beam would sweep across Jupiter. Given MAP's resolution—the sharpness of its focus—the planet should be indistinguishable from a mathematical point of microwave emission. "So we know what Jupiter should look like, given our mapping of the beam," says Page, "and we'll compare that to what it does look like, to see if the beam might have changed in some subtle way."

When the optics were complete, PCI, the contractor that built them, sent them to Goddard to join the radiometers and the thermometers and the waveguides. The latter were another precision-design problem, since these pieces of metal had to take signals from the antennas to the radiometers with minimal distortion; the waveguides for each pair of inputs for the differential measurements had to be exactly the same shape and exactly the same length—but all five pairs also had to intertwine for everything to fit, and they couldn't come too close to each other or they might contaminate each other's signal. "One of the hardest things to do," says Bennett, "was to take these radiometers and package them in the volume allowed and with all the constraints. A horrible problem. It was very tense." In doing a space mission, Bennett explains, you divide up the spacecraft's components so everyone can go off and work on them. Everyone has a physical space limit. The radiometer has to stop here, because this guy is building a propulsion system that's going to stop there. "A month later they come back and say, 'We need to come over three inches.' You go: 'Oh, no.'" Greg Tucker, from Brown University, led the effort on how to put this particularly nightmarish jigsaw puzzle together.

By the fall of 2000, all of MAP's components were down at Goddard and assembled with the rest of the spacecraft. Now it was time for the last round of testing. Al Kogut was in charge. "Basically, my job was to break the instrument on the ground. And if I couldn't break it, then it was gonna survive." For six months, the satellite was put through every test he could think of—plunged down to extreme cold, vibrated, bombarded with simulated microwave signals, blasted with sound waves from gigantic loudspeakers four feet across (which amounted to a vibration test at much higher frequency than you could do by just physically shaking the thing). In between tests, and sometimes during, the science team would check to make sure MAP was behaving itself. Given the small size of the team and the fact that many of the tests had to run twenty-four hours a day for days on end, just about everyone showed up to put in time.

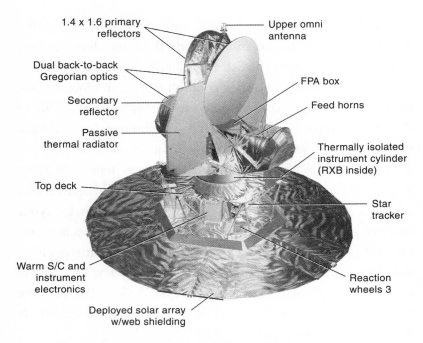

1.4 x 1.6 primary reflectors

Upper omni antenna

Dual back-to-back Gregorian optics

FPA box

Secondary reflector

Feed horns

Passive thermal radiator

Thermally isolated instrument cylinder (RXB inside)

Top deck

Star tracker

Warm S/C and instrument electronics

Reaction wheels 3

Deployed solar array w/web shielding

MAP, as fully deployed at L2. The satellite's mass is about one ton; it is about four meters tall and its solar arrays span five meters.

Inevitably, problems would emerge during this phase of the project as well. The biggest came after MAP had been fully assembled, and it came, not from any test at Goddard but from the National Reconnaissance Organization, which builds and operates spy satellites. They had some failures in a particular part—a power converter. "Unfortunately," says Bennett, "we had a lot of these parts on MAP—not hundreds, but quite a few, inside various electronic assemblies." The reconnaissance office was holding off on launches until they could procure new parts, but these wouldn't be available right away. That would have been a disaster for MAP, though. So the Goddard quality assurance office figured it was worth trying to fix them. The part sat inside a metal container, which evidently flexed when subjected to a vacuum and sometimes broke an internal connection. It took about a month just to figure out what to do. In the end, the QA folks decided the problem could be rectified if you epoxied a stiffener to the metal containers. So engineers took all the affected electronics boxes off the spacecraft and fixed the parts. "But you don't just put the boxes back on the spacecraft," says Bennett. "Now you've got to retest the boards and boxes." Ultimately, that setback cost the project three or four months. "Ed Weiler down at headquarters was pretty understanding about it. And I'm proud to say that if you subtract that time from the schedule, we came pretty close to meeting our deadline."

CHAPTER 9

Horse Race

By the time the fully integrated, fully tested satellite was ready to move down to Cape Canaveral, in April 2001, eight years had passed since Dave Wilkinson had first started thinking seriously about a post-COBE mission. During that time, the rest of the cosmic microwave background community didn't simply sit around waiting. In Europe, the Planck satellite, which really would "kill the problem dead," had been conceived, and organized, with the hope of a launch now delayed to 2007. Observers had also continued to set up antennas on the ground, and send them aloft in balloons, hoping to go beyond COBE in a less comprehensive but faster way than MAP could. As Lyman Page had once explained, an instrument on the ground or hanging from a balloon couldn't say anything definitive about the microwave background's characteristics over the entire sky, only over a relatively small patch of it.

But if you're looking for small enough ripples in that patch of sky, you can still be confident that they're at least broadly representative of what's going on everywhere. The biggest push after COBE was to find the acoustic peaks that reveal the physical attributes of Lemaître's "primordial atom." And if physicists could read them—read what amounted to the genome of the universe—they could finally understand what our infant cosmos looked like, in detail. They could move from the vague, ballpark description offered by COBE into what astrophysicists were beginning to call "precision cosmology." These telltale ripples should be about a degree across or smaller, so a balloon or a ground-based experiment, with a characteristic field of view of several degrees, should be able to pick them up. Such nonorbiting CMB detectors wouldn't be able to overlap with COBE's seven-degree scale, so they couldn't be directly calibrated to that satellite's definitive measurements. They wouldn't be able to average over the whole sky, so their precision wouldn't be all that high. And they would be plagued by many more sources of error than MAP would. But if the acoustic peaks were there, they should be detectable, and while MAP was taking shape, several physicists and astronomers had set out to look for them. Several had found them, long before MAP was even launched.

Among them was Lyman Page. His own graduate-thesis experiment had been a balloon-borne experiment that saw the ripples in the CMB just before COBE did. While the measurement wasn't definitive enough to report, it proved afterward to be perfectly consistent with the more reliable satellite data. It confirmed COBE, even though it came first. He knew firsthand, therefore, that there was no point just sitting around waiting for the next satellite to report in—even if he would end up working on it. So rather than sit around, Page plunged in. He worked with some folks at Goddard on a series of balloons that flew in the dry, frigid skies over Antarctica (the Medium Scale Anisotropy Measurement, or MSAM missions), starting in 1992. He worked on the Q-MAP balloon, which flew out of the National Scientific Balloon Facility starting in 1996.

He also worked—with Dave Wilkinson, Norm Jarosik, and Ed Wollack, among others—on the so-called Saskatoon experiment, operated from a prairie near that city in Saskatchewan, Canada, starting in the winter of 1993. What they found, after two years of careful observation, was that Dick Bond, George Efstathiou, and Jim Peebles and the other theorists who had talked about small-scale anisotropies seemed to be right. Looking at angles of a degree or so, Saskatoon saw the fluctuations in the CMB begin to depart from the scale invariance they showed at bigger angles. The chart that plotted the amplitude of fluctuations versus their size began to rise as they approached a degree. The so-called first peak, which would by its location measure the geometry of the universe—flat, positively curved, negatively curved—was evidently there, just as predicted.

Like the pre-COBE balloon experiment Page had worked on, this result was intriguing but not definitive enough to count as a discovery. But his next experiment, known as TOCO, did. This one was based on the ground, too, but since that ground was at 17,000 feet above sea level, at a site called Cerro Toco in Chile's Atacama Desert, there wasn't all that much atmosphere to look through. "The altitude takes a couple of days to get used to," admits Page, "and we had to live at 8000 feet and drive up to work on the experiment. But it's stunningly beautiful—like a moonscape but with snow-capped volcanoes popping off in the distance, and mountain ranges that take your breath away. You would look out and you'd see this perfect deep blue lake in the middle of this barren desert, and it would be full of pink flamingos. It's just one of these places that's magical."

Where Saskatoon had seen the spectrum of CMB fluctuations climbing out of the flat COBE plain, up to and just over what appeared to be the first acoustic peak, TOCO went all the way up to the peak and down the other side. As best they could determine, given the relatively small patch of sky they were looking at and the inherent margin of error in their equipment, the TOCO team put the highest point in the peak at an L of around 200. (L-numbers basically tell you how big the area of sky you're looking at is, by noting the number of pieces you've cut the sky into. The simplest way to think of it is just to remember that the number of patches is twice the L-number—so the dipole, with one hot and one cold spot, has an L of 1. The quadrupole, with four patches, has an L of 2, and so on.)

Based on the prior work of theorists, that meant that the geometry of the universe was indeed flat (or, given the uncertainties, at least close to being flat). COBE had confirmed one prediction of inflation—that on large scales, the amplitude of fluctuations would be even. Now TOCO had confirmed, albeit weakly, another. Page and his team published the results in the early fall of 1999, while MAP was still under construction. Given the immense importance of the result, one might have expected some press coverage—except that the TOCO team didn't call a press conference, or issue an all-points press release, or say anything about seeing God. The general public would learn about the result a month or so later, when James Glanz, a reporter for the *New York Times*, wrote an article with the headline, "Radiation Ripples from Big Bang Illuminate Geometry of Universe."

The news peg—the specific event that reporters use to justify why they're talking about this story now rather than last week or next month—was a result from an entirely different experiment, called BOOMERANG (Balloon Observations of Millimetric Extragalactic Radiation and Geophysics), a balloon-borne CMB detector that had circumnavigated the skies above Antarctica and examined about 3 percent of the sky. BOOMERANG had found the first peak in the power spectrum and was thus consistent with inflation. "Last month," continued the story, "scientists at Princeton University and the University of Pennsylvania . . . published similar results in *Astrophysical Journal Letters*." In fairness to Glanz, both publications were unheralded by press releases. TOCO's result was published in the *Astrophysical Journal*, while the BOOMERANG result appeared as a preprint on the Web. It took some digging to find both, and Glanz was the first major newspaper reporter to talk about either result. He also made it clear that the TOCO result had come first (although he mentioned it second—again, to suggest timeliness).

But from the way the article was written, the casual reader might easily have come away with the impression that BOOMERANG was first. That impression was reinforced six months later, when the results were formally published. This time there was a press release, which started off saying: "An international team of cosmologists has released the first detailed images of the universe in its infancy." The release was technically accurate, since TOCO had released a power spectrum, while BOOMERANG produced a map. "I think that's why they caught on," says Page. "They took their data and they made a really nice map that you could see. It was a great experiment. But it wasn't the first to see the peak."

Beyond that, the error bars in BOOMERANG's data were unconvincing. "You could be pretty confident they'd seen the first acoustic peak," says Al Kogut, "but where exactly that peak was, and how bright it was, you hardly knew." Six months later, when the BOOMERANG results were officially published, Glanz wrote again. This time, the headline read: "Clearest Picture of Infant Universe Sees It All and Questions It, Too." The question was, why hadn't BOOMERANG seen the second acoustic peak, whose height relative to the first should reveal the relative percentages of baryonic and dark matter in the early universe? The bolometers that the BOOMERANG team used, in preference to HEMT radiometers, should have been able to find the peak if it was there. If it wasn't, something was very wrong with cosmology. Glanz quoted David Spergel, among others: "That [the absence of peaks beyond the first] I think, would require a radical revision of cosmology as we know it." Indeed, says Max Tegmark, a theorist at the University of Pennsylvania: "About ten of us theorists wrote papers explaining why the second peak might not be there, inventing new physics. And then of course the next year they changed their mind."

It was a false alarm, just like the rocket measurements that had suggested a bump on the smooth curve of the CMB spectrum (unfortunately, the same astronomer, Andrew Lange of Caltech, was intimately involved with both experiments). The following year, the BOOMERANG team released a new, more accurate map of the sky, based on more data. This time, the first acoustic peak moved, and the second peak appeared. "You look at the two analyses, and they disagree," says Kogut. "And it just doesn't leave you with that comfortable feeling. If someone tells you something and then they come back a year later and say, 'Oops, did I say X? I meant Y.' What's there to say that they're not going to make another correction of comparable amplitude?" Again, the worry was the size of BOOMERANG's errors. "One of the problems with BOOMERANG is they don't know their beam patterns very well," says Kogut. The first time around, he says, they made a reasonable guess as to what the beam looked like. "Now, they have sort of a Mark 2 version of their beam, but they evidently still don't feel comfortable enough to publish it. They never put it out there for anyone to cross-check."

But other experiments would soon be reporting on their own attempts to detect acoustic peaks. And it wasn't only the CMB community that was trying to probe the fundamental structure and composition of the universe. Conventional astrophysicists were also looking at more recent epochs of the cosmos, to extend the revelations of early redshift surveys. The universe had, to everyone's initial surprise, a weblike structure with galaxies strung out along filaments and sheets that surrounded huge areas of relative emptiness. Now massive surveys of galactic redshifts like the 2DF survey, based in Australia, and the Sloan Digital Sky Survey, whose headquarters was in the basement of Peyton Hall at Princeton, were looking at broader chunks of the universe, to try and see that structure in greater expanse and detail. These were the end product of what had started out as ripples in the newborn universe, acted on by 10 billion years or so of gravity. Their precise configuration would be a clue as to what those ripples had looked like—and, pre-suming that the CMB community had their story straight, the two measurements, nearby and far away, should match in a sensible way.

At the same time, yet another set of observers had set out to measure a single cosmological number in an entirely different way. Until COBE came along, the only reason to believe in a flat universe was that it suited theoretical prejudice. Almost every survey of matter—dark matter included—suggested that there was much too little, by a factor of order ten, to slow the expansion of the universe. Omega, as far as anyone could demonstrate, was 0.1—maybe as much as 0.3 at the very most. For reasons laid out earlier, though, that was so close to 1 that theorists insisted that the actual number must really be 1. You could make up the difference by positing some huge amount of dark matter that hadn't yet been detected.

Or you could bring back Einstein's cosmological constant. Since matter and energy were two sides of a single coin, the geometry of space could as easily be warped by the latter as by the former. So if, as Einstein had reluctantly and temporarily presumed, a cosmoswide, outward-pressing force really did exist, its energy would contribute to omega. If the energy content were large enough, it could boost omega all the way from 0.3 to 1.0. The cosmological constant, generally denoted by the Greek letter lambda, could flatten the universe. Physicists didn't much like this idea, though. Long after Einstein had abandoned his most distasteful proposition in the happy discovery of an expanding universe, particle theorists had found reasons of their own for suggesting the existence of a cosmological constant. Quantum mechanics—specifically, Heisenberg's celebrated uncertainty principle—suggested that on very short timescales, pairs of particles should spontaneously appear from the vacuum, then annihilate one another before their existence could ever be noted.

This continual bubbling of so-called virtual particles should in principle charge empty space with energy. The only problem was that the amount of energy involved should be prodigious. Einstein's cosmological constant would have been just powerful enough to counter the effect of gravity. But the energy from virtual particles should be about 1,000,000,000,000,000,000,000,000,000,000,000,000,000,000,000,000,000,000,000,000,000,000,000,000,000,000,000,000,000,000,000,000,000,000,000,000,000,000,000 times stronger than that. If that were actually the case, the universe would have flown apart so quickly after the Big Bang that electrons would never have had the chance to combine with nuclei. Atoms wouldn't exist, to say nothing of stars, galaxies, planets, and human beings. Invoking the cosmological constant to flatten the universe, then, was a dangerous proposition, since you had to explain why it was so remarkably close to zero, but not quite (essentially the same problem, in fact, as arguing how omega could be so close to, but not exactly, 1).

"The cosmological constant," Princeton astrophysicist Ed Turner once wrote in a review article, "is an idea whose time has come . . . and gone . . . and come . . . and so on." In its most recent visitation, it was being invoked as a way to boost omega to 1, and flatten the universe. For reasons noted above, this was a tough sell—especially since there was no direct evidence for it. In 1998, though, not one but two independent teams of astrophysicists stumbled on that evidence. Brian Schmidt, who had gotten involved with this area of research under the tutelage of Robert Kirshner at Harvard, headed one group out of the Mt. Stromlo Observatory in Australia; Saul Perlmutter, of Lawrence Berkeley Laboratory, led the other. Both groups were trying to nail down the value of omega by gauging how fast the universe's expansion was slowing down. If omega were low—if there weren't a very high density of matter in the universe, and not much gravity as a result—then the expansion should be proceeding pretty much unchecked. If omega were high, on the other hand, the constant drag of gravity should mean the rate of expansion should be lower now than it was in the past, as the universe decelerated.

Both groups set out to measure the slowdown, and gauge omega, by looking at Type 1A supernovas, a particularly bright variety of exploding stars whose light output was highly consistent. In a minimally decelerating universe, the relationship between redshift (and therefore recession speed) and brightness should be reasonably steady; in a fast-decelerating universe, the more distant supernovas should look brighter than expected. But what both groups found was that in fact they looked dimmer. Adam Riess, working with Schmidt's group, first from a post at Berkeley and later from the Space Telescope Sciences Institute, remembers thinking that this couldn't be right. There must be, he thought, a bug in his software. There was no bug, however. Perlmutter's group, equally baffled, went through its own exhaustive reanalysis of the data. In the end, both groups reported, more or less simultaneously, that the expansion of the universe wasn't slowing down at all. It was speeding up.

These observations were startling and important enough that *Science* magazine went on to declare the accelerating universe the "Breakthrough of the Year" for 1998. (When the MAP team suggested in its proposal a couple of years earlier that the satellite could measure the cosmological constant, other astrophysicists had acted like they were crazy for even mentioning such a fringe phenomenon.)

For cosmologists, it offered an ideal way to complete the omega jigsaw puzzle. Prejudice, COBE's scale-invariant spectrum of fluctuations, and, eventually, TOCO's detection of the first acoustic peak would all suggest that the universe was flat; that omega was equal to 1. (If omega had turned out to be 1 as the result of matter alone, the expansion of the universe would have slowed forever, but never quite stopped. Since the biggest contribution comes instead from a form of energy that acts in opposition to gravity, it's now more certain than ever that the universe will expand forever, and at an ever-increasing rate.)

Not only was the universe flat, but BOOMERANG's subsequent detection of the second acoustic peak would set the contribution of matter at about -0.3 or 0.05 worth of ordinary baryonic matter, the rest cold dark matter in some form. That jibed nicely with calculations based on the motions and rotations of galaxies, and with the amounts of light elements formed in the Big Bang. And now, calculations were showing that the observed acceleration would require a cosmological constant whose contribution to omega would be about 0.7. In short, all known sources of curvature were adding up to an omega of 1. The universe appeared to be flat, just as the theorists always wanted it to be. Inflation had evidently passed its second major test, and the consensus was now even stronger for a Standard Model that included inflation, a cosmological constant (which quickly acquired the catchier name "dark energy"), plenty of dark matter, and a dusting of baryons.

Still, even as other CMB experiments broadly confirmed TOCO's and BOOMERANG's results over the next few years, none could really be considered definitive. MAXIMA (Millimeter Anisotropy Experiment Imaging Array), which was another balloon that flew over North America; and DASI (Degree Angular Scale Interferometer), a ground-based experiment at the South Pole, all reported that they saw the first peak, and several of the experiments claimed to see the second as well. But they didn't all see the peaks in precisely the same places. That might be fine, since all of them, like BOOMERANG, suffered from a fair amount of imprecision. But you had to take the observers' word for how great that imprecision was. "To be very frank," says Tegmark, "what makes me very nervous about some of these experiments is that they never have made their data public. This to me is one of the most important things about MAP. It will make its data public. I'm really bothered by this fly-by-night stuff, where they make big press releases and say here is the final answer. And they say the data's great, trust us. But we're not going to show it to you."

One reason MAP would publish its raw data was that this was NASA's policy. The public paid for this expensive piece of hardware, so the public should get to see everything, at least after getting first crack at analysis. COBE worked the same way. But releasing the data was also something the MAP team would have done, anyway. "We've made the data public on just about every experiment I've ever worked on," says Lyman Page, "along with the code we used to analyze it, and the analysis, and everything. TOCO, Saskatoon, all of them." You do run the risk that someone else will come up with a better code and do a better analysis, he says, but you also get lots of other people double-checking your work, so there's less chance of having to make an embarrassing retraction. "It's just the right thing to do."

Chuck Bennett emphasizes that all of these groups have done terrific, important science. They've pushed the envelope of what's possible to do from the ground. But given the vagueness of their results, none of them has weakened the need for a satellite like MAP. "In a broad, general sense, there's evidence that the universe is flat, but we can't say with great precision exactly how flat, to within 10 percent or so." MAXIMA, for example is consistent with a flat universe, he says, but its best fit is an open universe. BOOMERANG's best fit is to a flat universe. "These points should be on top of each other, and they're not. What that means to me is that the errors are larger than what they're saying. They're all looking at the same sky out there. Which is to say, we don't need more sensitivity. MAP is going to resolve it."

CHAPTER 10

Launch

Down at Cape Canaveral, the pressure on Chuck Bennett and the rest of the team was, if anything, more intense than it had been at any time during the preceding six years. "I hated being in Florida," he says bluntly. "It was extremely hot, and I don't like hot weather. And then you had all this pressure and all these different people to deal with. They were all very friendly and helpful, and good to work with. But there was a lot going on at once. You had the Air Force to deal with [because MAP would be launched, not from the Kennedy Space Center, but from the Cape Canaveral Air Force Station a few miles away] and Boeing [which made the rocket] and the Kennedy Center people [because it was a NASA project, they were in charge of the launch despite its location]." He was also getting preliminary calls from journalists now and helping to figure out how the press briefings should be handled before, during, and after the launch, and what guests should be invited. "There were just a hundred thousand things going on. My mind was going in all directions at once."

Just as in every other phase of the project, new problems kept showing up—glitches in the star trackers, light peeking through the solar shield, electrical interference from somewhere or other. "It's a completely different world," says Cliff Jackson. "In this phase, you don't get to sit around for a month studying how to fix things. You've got a whole marching army, everything's sitting there. You've got to decide very quickly whether a problem really needs to be fixed, and how to do it. You're always asking yourself, 'Is it good enough? Do I really know? Am I sure?'"

Originally, the team had hoped to be sure by November 2000. NASA's requirement for the MIDEX program had been to launch by March 2001. In the end, MAP was on the pad at the Cape Canaveral Air Force Base in June 2001. In order to use as little fuel as possible in getting to L2, MAP was going to use the Moon's gravity as a kind of slingshot, falling in toward it on a skimming trajectory that would whip it out into deep space. To get into the ideal position to do that, the satellite would actually loop out toward the Moon, then back to and around Earth, then repeat the maneuver; the gravity boost would happen on the third time out. This complicated series of loops required launches at particular times of the month. On June 30, at 3:47 in the afternoon, the launch window would open for just twenty minutes and would reopen at the same time for the next four days only. Then they'd have to wait another two weeks. When that first opening came, a few of the MAP team members would be at the Cape, with their families. There wasn't anything left for them to do, so they could simply show up for the festivities and enjoy themselves—as much as you can when several years' worth of intense work is sitting on hundreds of thousands of pounds of explosive fuel.

But most of the team was working. Somebody had to monitor the spacecraft during the launch, for example. As soon as it left the pad in Florida, the satellite would revert from the hands of NASA controllers on the Cape to a control room at Goddard, where Norm Jarosik, Steve Meyer, Lyman Page, Greg Tucker, Gary Hinshaw, and Ed Wollack would be sitting at computer terminals, anxiously performing diagnostic tests on the radiometers, the optics, the electronics, and everything else that had to survive, first the launch itself and then the transition, within only an hour, from the heat and humidity of a Florida summer afternoon to the frigid vacuum of space.

The Cape was, therefore, a more festive place to be, for those few members of the team who could afford to be festive. The night before the first launch opportunity, David Spergel, Ned Wright, and Dave Wilkinson hosted a party in a slightly seedy, faux-glamorous meeting room at the Cocoa Beach, Florida, Holiday Inn. It resembled nothing so much as a low-budget wedding reception: men and women of all ages sat around big round tables, getting up occasionally to buy a margarita at the cash bar or go through the hors d'oeuvres line for another platterful of lukewarm fried shrimp or stuffed mushrooms. Children tugged at their parents' sleeves, and were finally sent out into the hallway to run back and forth, shrieking with pleasure at being liberated. Mood music played in the background, but it was impossible to identify over the chatter and laughter.

As Spergel went from table to table to visit with the guests— not just scientists, and engineers, but aerospace contractors who had worked on MAP, and even some of the Princeton machine-shop guys who had helped Norm and Dave Wilkinson with the instruments—Laura Spergel sat with the three kids, her parents-in-law and David's sister and her family, and Laura's two brothers and their families. Lyman Page's parents and sister were there ("I'm the real Lyman Page," Page senior would tell people, with a broad smile); his wife, Lisa, was out on the beach with their three boys. Norm Jarosik's parents showed up as well. Dave Wilkinson's extended family—his wife, Eunice, her two grown daughters from a previous marriage, his son from a previous marriage, and assorted grandkids—took up an entire table. Jim Peebles, Dave Wilkinson's partner from the original, scooped CMB-detection experiment, was there with his wife, Allison. Ned Wright stayed for a little while, but his family couldn't resist the lure of the beach, with water far warmer than the Pacific they were used to.

The mood was very much like a wedding, too, said Spergel. "We've got families here, and colleagues, and friends, and none of the groups really know each other, but everyone's happy and excited." As the evening progressed, he began to use a different analogy. "For me, I'm thinking that this project is like childbirth—a Caesarean section. I had a role in the conception. The birth about to happen, but I can't do anything to help or hinder it; I can only be nervous and watch. If it goes badly, I'm going to be devastated. If it goes well, I'm gonna have a lot of work to do and a lot of sleepless nights for the next year or two."

The next morning, guests began gathering at the Kennedy Space Center's Visitor Center, from which buses would depart for the launch site, a half-dozen miles down the Cape. The place was packed with tourists, as always—gazing at rockets from the 1950s and 1960s in the Rocket Garden (a golden eagle had built its nest atop one of them); snacking at the Launch Pad or Orbit restaurant; catching a show at the IMAX theater; buying souvenirs (dehydrated Astronaut Ice Cream was a big seller) at the

Space Shop, listening to an alfresco talk by a Real Astronaut. Almost none of them had a clue that the most important cosmology mission in a decade was about to take off, nor that some of the scientists who had made it possible were circulating quietly among them.

Just after lunch, a half-dozen yellow school buses pulled up in front of the center's Accreditation Office, and all of the MAP guests, along with several of the scientists, piled on for the ten-minute drive to the air force base. Chuck Bennett, meanwhile, along with Cliff Jackson and Liz Citrin (the woman who had taken over from Rich Day as project manager at Goddard when Day was offered a promotion too fat to refuse), headed for Mission Control at the space center, where they'd literally be plugged into what was happening. They'd have to give the final OK that the satellite was ready before the launch could actually happen. "If you see something wrong," says Bennett, "you have to say 'hold, hold, hold,' three times, so they're sure you mean it. I had the authority to stop the countdown." At times, the Goddard folks were listening to ten different voices at once over their headphones reporting on the weather, on the rocket, on the satellite, on the boat and plane traffic in the area, and more. Keeping track of all these voices is generally impossible for the inexperienced, but Bennett was clearly doing it. "The people in Mission Control were amazed," he says. "They asked how I could do it. 'No problem,' I replied. I was a ham radio operator."

As most of the MAP team were sweating through the launch at the Cape and at Goddard, the bus riders unloaded in a parking lot next to a couple of utterly nondescript military buildings that looked vaguely like the large rural regional high schools built around 1960, and walked through an opening in a chain-link fence to a set of bleachers—think high school again—facing what at first looked like just a line of trees. A road headed off through the trees, though, and if you positioned yourself right, you looked down that road and saw what looked like a toy rocket, sitting on a pad. David Spergel wasn't there, though, nor was he at Mission Control. A British television crew was producing a documentary and wanted to film somebody during the actual launch. "I was about the only one on the team," he says, "who had nothing useful to contribute, so they picked me." As a result, Spergel and his family were out on a pier, where the rising rocket would be easy to see. "The kids were shouting out the countdown," he says. "It was all a lot of fun."

It would be another hour before the launch could happen, at the earliest. Luckily, NASA had provided a snack wagon, so some of the world's most eminent astrophysicists stood in line in the hot sun to buy hot dogs and ice cream for the kids and grandkids. A loudspeaker squawked out the chatter from the NASA ground controllers—small planes and boats were wandering close to the forbidden zone that surrounded the pad; some clouds that could cause trouble were moving in from the west; high-altitude winds seemed OK for now; no obvious troubles with the rocket or the probe. Anyone who's paid the slightest attention to the space program knows that a launch can be canceled for any one of a thousand reasons, from active thunderstorms to a funny signal from a sensor. A successful launch can be followed immediately by a disaster, of which the *Challenger* explosion is merely the most tragic and famous of hundreds. Even if immediate disaster is averted, the satellite has to go into its proper orbit, it has to turn on successfully, and it has to function after that happens.

So when the countdown went to zero and a cloud of smoke and water vapor billowed from the base of the Delta rocket,

followed by a flame, followed by the slow rise from the pad—
followed, finally, by a roar, the final sensory message to reach
the bleachers—everyone cheered. "I kept thinking, 'We really
made it, I can't believe it,'" remembers Bennett. "'No more
wrapping of insulating blankets, no more testing." He couldn't
relax, of course, and neither could the other science team members; this was just another, though admittedly important, milestone. Even as the rocket gathered speed, trailing flame and
smoke and finally disappearing into the hazy Florida sky, their
exultation was tempered by anxiety—about whether the spacecraft would go into orbit, whether the instruments would turn
on properly whether a heated blade would cut through a Kevlar
rope to free up the solar panels successfully, and make it to L2,
and a million other things. As soon as the rocket was off the
ground, Bennett drove over to another operations center where
Al Kogut and Michele Limon watched the satellite turn on.

Up at Goddard, meanwhile, the rest of the team was too busy
working to worry much. "Everyone wanted to be the first to see
the signal that MAP was working OK," says Page. "But the truth
is, I don't remember who it was." The signal came about an hour
after launch, after the satellite had detached from the rocket and
gone into Earth's orbit in preparation for its next maneuver.
"When the receiver came on at the right level and sent back its
first signal—I think we detected the Moon—it was just such a
relief. I think we cheered, but I can't even remember." That detection meant that MAP had already survived not only the
launch and the exposure to space, but also passed another potentially lethal test. Given the orbit and the way MAP was launched,
there was no way to avoid pointing the telescope directly at the
Sun for a moment. Such an operation could literally have burned
out the optics. "You can fry an ant with a magnifying glass,"

says Page, "and when you focus the Sun down your receivers
. . . well, we spent the better part of a year trying to figure out
how dangerous this would be for us." In the end, they roughened
the metal surfaces of the optics enough that they'd diffuse the
sunlight without affecting the reflection of the longer-wave-
length microwaves.

Ed Wollack and Norm Jarosik, meanwhile, were happy the
radiometers were working, but they needed to know precisely
how they were working. How did the response of each one differ
from what they expected from previous testing, and why? "All
of the amplifiers are kind of like Marion's children," says Wollack, "and each one kind of has its own personality—which I'm
pretty familiar with by now." Thanks to the thermometers and
other sensors on board, the team could also measure the satellite's mechanical health. It was predictable, and predicted, that
as MAP orbited into and out of Earth's shadow, everything
would heat up and cool down, making the satellite expand and
contract. "Everything goes crazy," says Wollack, "which is fine.
But there was also something we didn't understand. We saw that
there was some sort of extra torque on the satellite—it was twist-
ing in a way we didn't expect." It wasn't a huge problem, but
the team was obsessive about tracking down every last detail.

So they made a list of all the possible causes. It was a long
one, but one by one they thought about each item, and ruled it
out. "We hate it when that happens," says Wollack. "It's a lot
better when you have one explanation left over." So they began
making new lists. The torque was periodic, so maybe the thrusters had deposited some kind of gunk on the reflectors and they
were causing it. "So we did the numbers on that, and it wasn't
possible." They looked at the data harder and realized that the
torque was pushing one way, then the other. It was, they

thought, as if the reflector were covered with ice. One side would get heated by reflected light from the Earth, and some of the ice would blow off, and the satellite would recoil. Then the other side would get heated, and it would recoil in the other direction. "We ran the numbers for that," he says, "and it was like, 'Wow, you can make something like that work.'"

Except there was no way to get enough ice on the reflectors, and where would it come from, anyway? It took Wollack and Cliff Jackson and several others a few days to figure out that the ice theory was right; it was just the location that was wrong. Ice had formed on the blankets on the back of the solar panels, not the reflectors. It had come from the Florida air—humidity soaked up by the satellite's insulating blankets, then heated by the Sun, driven off and refrozen on the solar panels, then driven off again in little spurts. "It basically turns into thruster fuel." They knew for sure that this is what was happening when they charted its future course: every time some ice blew off, there was less available for the next time around, so the force should be less. And it was. "Somebody had probably seen something like this before," says Wollack, "but they weren't on the team."

Finally, the satellite's thrusters boosted it out of Earth's orbit, on the way to the Moon to begin maneuvering for the first of three Earth-Moon loops that would position the craft for its final gravity assist. Chuck Bennett fretted his way through that entire phase of the mission. "You're coming around this loop," he says, "and right at this instant you're supposed to fire, and then get so much force out of that thruster. What if it's a little more, a little less? What can you tolerate? There's plenty of possibility for completely losing your mission in doing that stuff." People kept asking, 'What if you don't make it to L2? Would that be OK?' They wanted the mission to be salvageable, so they could tell everyone it was just fine. I said, 'No, it would not be OK. That would be a mission failure.'"

The pressure was finally too much, and on July 4, confident that MAP was on its way at last, Chuck Bennett allowed himself a rest day. He was lying in bed when he started feeling intense pain in his lower back. He tried a hot shower, but that didn't help. He tried exercising on his elliptical trainer, a treadmill-type machine. That made it worse. Eventually he woke his wife, Renee, who called the doctor, who thought he might be having a heart attack. It was now two or three in the morning. Nobody was on the road, and the usually conscientious Bennett ordered Renee to ignore the traffic signals. They ran every red light on the way to the hospital. Bennett's heart was fine, as it turned out, but his gall bladder was not. Ordinarily, a failing gallbladder can be removed with a laproscope, a thin tube with a fiber-optic scope and a little scalpel on the end.

Not this time: once inside, the surgeon saw that gangrene had set in. Bennett had been so preoccupied with the satellite that only in retrospect did he recall feeling a few twinges during the past several months. Suddenly, a simple operation had turned into major abdominal surgery—and if the doctor hadn't gotten every piece of the organ out, a very nasty bacterium could have been sent coursing through his entire system. "Once that happens," says Bennett, "you have a 50 percent chance of survival." The surgery left him feeling totally wiped out. "My wife said she knew I wasn't feeling well when I didn't ask about MAP for a whole twenty-four hours." The doctor said he needed to stay out of work for a month, but he was back at his desk in a couple of weeks. "So it almost killed me, but I got this thing launched."

It would take a month for MAP to complete its three loops around the Moon, and the probe wouldn't reach L2 until the end of September 2001, about three months after launch. But that didn't mean the science team had a three-month rest. The satellite was taking data the whole time, spinning once every two minutes, taking fifty snapshots of the microwave sky every second, each a differential measurement of two points 140 degrees apart, just as though it had begun its official science run. The point wasn't to do science, though, but rather to measure the instruments, to compare their actual performance with predictions based on Princeton's and Goddard's exhaustive testing on the ground. Lyman Page would watch as Jupiter drifted across the satellite's beam, and he carefully mapped the antennas' response, just as he had with a bright light on the math building in Princeton. Norm Jarosik and Ed Wollack monitored the receivers' response. Gary Hinshaw made maps of the sky and compared them with his simulations.

And pretty much everything worked as planned, within acceptable margins of error. That didn't mean there weren't surprises; simply that, just as with the ice on the solar panels, the team was finding ways to deal with them. Every so often, a solar flare would erupt, sending a gust of charged particles toward Earth, and thus toward MAP. "Whenever we hear 'flare' on the radio," says Spergel, "we see it as a minuscule rise in the satellite's temperature. Once we had to go into safe mode, during a bad one." But basically, MAP, sitting in the cold and darkness of L2 and armed with technology that's twenty years newer than COBE's, is performing spectacularly well. "The mapmaking software is working so well," says Page, "that when the first data run ends on March 30, they'll have an actual map of the sky, albeit a preliminary one, the very next day."

Only one real shock would hit the team during the six-month data run, and that was a pleasant one. "I got a call one day in the fall," says Spergel, "from the MacArthur Foundation." He'd been asked to comment in the past on candidates for this prestigious and lucrative award, but he became suspicious when he was patched into a speakerphone in what was obviously a meeting in progress. "I figured there were two possibilities," he says. "Either they wanted a comment on a finalist, or they wanted to give me the award." The head of the foundation asked Spergel if he knew any MacArthur fellows personally. He ticked off a few names. "Actually," said the voice on the phone, "you know one more. You." The MacArthur, unlike the Nobel, isn't given for a single achievement, but rather for a track record of creative thinking.

His work on MAP alone wouldn't have qualified, but Spergel had lately done another piece of spectacular scientific work in a completely unrelated field. A group of colleagues at Princeton were competing to build a different sort of satellite, one that would be able to take images of Earthlike planets orbiting distant stars. The big problem here is that planets are very close to stars, and are also very dim compared with stars; blotting out a star's light so the planet shines through turns out to be an extraordinarily hard task. But Spergel, recruited to be a member of the team designing this satellite, took home a book on optics and taught himself this branch of physics, which he'd never studied, on his own. Out of that study came a design for a new kind of telescope that could blot out starlight in a radically new and potentially highly efficient way. Best of all, the idea was so amazingly simple that a colleague who specializes in optical engineering said, "It's so obvious in retrospect that I'm kicking myself that I didn't think of it." (The idea basically involves putting a mask with a cat's-eye-shaped opening over the end of the telescope, which, for complicated reasons, shunts all the starlight away from the center of the image.) "When I heard about it," says Ed Turner, Spergel's colleague in the Princeton astrophysics department, "my jaw dropped."

Naturally, most people wanted to know what the first thing was he'd done with the money, which generally runs to about a half-million dollars. "The truth," he says, "is that I bought a really nice foosball table." The second thing, though, was that his wife, Laura, could quit a well-paid but unsatisfying job at the state's department of health and move on to work on issues of bioterrorism, which interested her a lot more. "So while a lot of people say the MacArthur had a huge effect on their careers, and while it may have that effect on mine," he says, "it's really changed Laura's life more so far."

CHAPTER $\boxed{11}$

Deepening Mystery

By the first weekend of June 2002, the chilly weather that had persisted for most of the spring had finally given way to a heat that would soon become oppressive. Every year at this time, Princeton alumni return to campus for a three-day reunion party that's unsurpassed by any university in the country for drinking, carousing, and general silliness. Unlike, say, Harvard, which makes a fuss only over the milestone-anniversary classes—the tenth, the twenty-fifth, the fiftieth—Princeton entertains a contingent from just about every class. Parties go on for most of the night in tents set up in the university's fields and courtyards, with separate activities for kids. On Saturday afternoon, everyone gathers for a parade through campus, the most memorable feature of which is the costumes the alumni wear, all of them in some variation of orange and black, with or without the school's trademark tiger as part of the design. The costumes are sometimes merely loud, but sometimes they're downright absurd. This year, David Spergel's class went for the absurd: to celebrate its twentieth reunion, the Princeton Class of 1982 would be marching in fuzzy orange and black shorts and orange and black striped shirts.

As he was walking, however, Spergel's mind was approximately 14 billion light-years away. Within a few days, the MAP team would be getting its second set of processed data. For obscure reasons, this second pass at the first six months of observations was called "pass one." The first pass was "pass zero." The team had begun constructing maps of the microwave sky as soon as the first run ended back in March, but those early maps were mostly for practice; the maps of pass one will be a lot more reliable. To the eye of a connoisseur, they'll be more than that. Rather than limit themselves to a single computer screen, the team enlisted the help of Ben Shedd, a computer-graphics expert who had constructed a gigantic, high-resolution screen in the university's computer sciences department. More than once, they'd marched uphill to the projection room, gotten Shedd to load their data on his computer, and then made him leave the room. "It's just mind-boggling to look at it," says Lyman Page. "It's like you're walking down a drab corridor, and you open a door, and suddenly there's this sparkling, spectacular vista like nothing you've ever seen before."

Even on pass zero, though, Spergel could see that something peculiar was going on. Ten years back, when COBE's differential microwave radiometer came in with George Smoot's vision of God, it appeared to vindicate a key prediction of inflation. Inflationary theory said that the variation between any two spots on the sky should be the same no matter how wide their separation—a pattern that reflected the characteristics of the primordial energy field that triggered the breakneck expansion known as inflation. Within its margin of experimental error, that's what the DMR had found—or so everyone thought.

During a chilly week in April, though, as Princeton bloomed, David Spergel had realized that MAP was seeing fluctuations that didn't match what inflation predicted, after all. The pattern of scale invariance that COBE had evidently confirmed was there at small angles. But on the very largest scales, in patches of the sky spanning 60 degrees and more, the variations looked weaker than they did in smaller patches. In fact, they were almost nonexistent. "I took this to the group, of course," he said a few days after the discovery. "Their first response was, 'Why didn't we see this in the COBE data?' Ned Wright went and pulled the original paper . . . and there it was." Nobody had noticed this before—not even Spergel. The reason was that the COBE data were presented in two different ways. The first was purely as a power spectrum: the strength, or amplitude, of fluctuations on the Y-axis versus their angular separation on the X. Due to the way the power spectrum is calculated, large scale deviations simply don't show up. When you look at the data in a different way, though, and measure the degree to which two points on the sky are correlated in amplitude depending on angular separation, the discrepancy is glaring. "I can't believe," says Spergel, "that we missed it. I think it's because so much information was coming out at once, and people were distracted by the power spectrum, which seemed to support inflation."

This deviation from scale invariance didn't prove inflation was wrong; it could show up purely by chance. It was unlikely— the odds were something like 1500 to 1 against it happening— but that didn't mean it was impossible. Ideally, you'd want to run the experiment again to check, since the odds against such a deficit showing up twice were literally astronomical. Unfortunately, you couldn't do it. It takes only a handful of patches this large to cover the entire sky, which means there aren't that

many independent measurements you can make. It's like flipping
a coin: if you only get ten flips, you'll sometimes get eight or
nine heads, even if it isn't very often. If you get a thousand flips,
you'll essentially never get eight hundred or nine hundred heads.
With only a dozen flips, the result isn't airtight. Or think about
the fact that the Moon and the Sun appear to be almost precisely
the same size on the sky, even though the Sun is much bigger and
farther away. It's why we have solar eclipses. Yet while it might
be tempting to wonder why it happened to turn out this way,
said Spergel, "there's no reason for it, and the odds are against
it happening. Nevertheless, we don't think it means anything."

But another reason the COBE measurements of large-scale
weakness weren't taken too seriously, said Spergel, is that humans have a built-in tendency to ignore data that don't fit their
preconceived models of how the universe is supposed to work.
In 1054, he pointed out, a supernova—an exploding star—burst
into sight. Chinese astronomers duly noted the appearance of
this "guest star," but it appears nowhere in European chronicles,
even though it outshone every other star in the sky. "Europeans
knew the heavens couldn't change, so they didn't take it seriously," he said. There's also, he acknowledged, a tendency in the
opposite direction as well.

Now MAP, with its tighter error bars, had found the same
thing COBE did. And this time, someone noticed. "We have no
theory," Spergel said, "as to what it means yet. It could be that
simple inflation is only part of the story in the very early universe. Maybe there were multiple, overlapping inflationary
fields. Or maybe . . ." He paused to consider how far out on a
limb it's appropriate to go this early in the game. "One thing
that could produce an effect like this is a small universe." How
small is small? "Maybe . . . half the size of the visible universe."

He knows that if this turns out to be true, it would be shock-ing. It would mean that the most distant galaxies we can see aren't distant at all; they're nearby galaxies whose light has trav-eled once around the universe and come back to hit us a second time. It would mean that among the faint images of those far-away galaxies, captured by the Hubble telescope in space or the giant Keck or Subaru or Gemini or VLT telescopes on the ground, may be the Milky Way itself—a baby picture of our home galaxy, whose light has come to us the long way around.

Despite the science-fictional notion of such a universe, it isn't as preposterous as it sounds. In a closed universe—the version that's analogous to the surface of a sphere—it's easy to see how it would work: a beam of light from a galaxy would simply go once around, like a line of latitude or longitude, and come back to where it started. That would work equally well in three dimensions, as long as the cosmos was smaller in light-years than the age of the universe in years (otherwise the beam wouldn't have made it once around yet). You still have to accept the fact that a three-dimensional cosmos can curve back on itself but that, unlike an analogy to a sphere, it doesn't need to have a fourth spatial dimension around which to curve. But once you do that, it makes sense.

With a flat universe, though, there is no curvature, and by now it's pretty clear, thanks to all sorts of independent experi-ments—involving not just the microwave background, but also those that show evidence of dark matter and dark energy—that the universe really is flat, or very, very close to it. Unlike the case of the surface of a sphere, the proper analogy would be a flat sheet of paper. So how do you deal with the edges?

The surprising answer comes from topology, the field of math-ematics that deals with the connectedness of shapes and surfaces.

Topology explains, for example, why a coffee cup and a dough-nut have essentially the same shape (each one basically consists of a hole and some other stuff attached; though their appear-ances are different, their connectedness—the number of direc-tions you can go in and return to your starting point—is identi-cal). Topology also explains how a flat universe can exist without curvature, without edges, and yet can still be finite in extent. "Say I have a rectangular sheet of paper," says Spergel, "and I tape two opposite edges together. Now I have a cylinder, and I can travel around the outside and come back to my starting point, without encountering an edge." Bend the cylinder around so the two open ends are touching, and tape them together: now you have a doughnut, and you can travel in any direction with-out reaching the edge. "It's like the screen on a game of Pac Man," says Spergel. "If you travel beyond what looks like the right edge, you immediately appear coming in from the left."

The catch is that when you bend a piece of paper, it's no longer flat: you have to bend it through a third dimension to create the doughnut. Bending a flat universe isn't allowed. But—and here is another one of those statements that turn common sense on its head—the topology of spacetime permits a flat, finite, three-dimensional universe to connect to itself without "bending" through a higher dimension. That being the case, it's possible in principle that light could travel all the way around the universe and return to its starting point. Just how it would do so depends on precisely how the universe is connected to itself. It can con-nect like the sheet of paper mentioned above, in which case there's only one way to go. It can connect like a doughnut, mean-ing you could look in two different directions—the long way around the perimeter, or the short way, around and through the hole—and see yourself in the distance. It can even connect like a more complex object, such as a "doughnut" with two or even three holes, which would make the number and pattern of multiple images correspondingly more complex.

In all of these cases, however, multiple images can occur only under one condition: that light has had enough time since the Big Bang to make a complete circuit of the universe. If the cosmos is infinitely large, or even just slightly larger than the horizon—15 billion light-years around, say, with a Big Bang only 13 billion years in the past—then you won't see any extra images at all. They haven't gotten here yet. If, on the other hand, the universe is small compared with its age, light could have made several circuits since the Big Bang. You could in principle see the image of the Milky Way in the distance, and again in the far distance, and again in the really really far distance. It sounds crazy, but in a way the alternative is crazier. "If the universe is infinite," says Spergel, "then I'm saying this sentence an infinite number of times, except that sometimes I'm saying it backwards and sometimes in French and sometimes in French with a Spanish accent." Philosophically, that notion is troubling, to say the least.

Even so, he said, "prejudice at the moment favors infinite universes. But historically there have been periods where people like Einstein thought that the universe ought to be finite." For that reason, he had always planned to look for evidence for a finite and (because anything else would be undetectable) small universe. It would show up as a distinctive pattern of unusually hot and cool spots, as two or more versions of the microwave background repeated or even overlapped in different parts of the sky. "I'd say the odds are against finding it," he admitted early on, "but the importance of finding it if it's there is so great that it's worth spending time on doing the analysis." Now, the weakness of the microwave signal at the largest scales on the sky had given him a concrete reason to look, since a finite universe would lead to just such a weakness. "I've just revived my topology code to look for a small universe," he had said on that cold day in April. "Last night, for two hours, I was convinced I'd found it. Then I realized there was an error in the software." And ultimately, it would turn out that Pass Zero would be too raw to show the telltale signature of a finite, small universe.

But by early June, Spergel had gotten excited about the idea once again, thanks in part to an entirely different line of evidence that had emerged from Pass Zero and in part to a new result from a cosmic microwave background experiment prerched on an arid plateau called Llano de Chajnantor in the Chilean Andes. The Cosmic Background Imager, a joint project of the National Science Foundation, the California Institute of Technology, and the Canadian Institute for Advanced Research, had been monitoring three patches of the southern sky, each about seventy times the size of the full Moon. Like John Carlstrom's DASI experiment at the South Pole, CBI is an interferometer—in this case, a suite of thirteen microwave antennas that operate in concert to probe much smaller fluctuations than a single telescope could do. But CBI is more sensitive: it can discern variations in the primordial radiation as small as three minutes of arc across—about 140 times smaller than what COBE could see. (It also beats DASI for altitude: at nearly 17,000 feet, CBI is a mile and a half higher than the South Pole, and its air is consequently so much thinner that astronomers carry individual oxygen bottles with them when they go observing.)

At that level, CBI detects fluctuations as small as individual clusters of galaxies, and where the BOOMERANG and TOCO and other experiments had managed to detect the first and second acoustic peaks in the fluctuation spectrum—the telltale bumps that represent overlapping pressure waves in the early universe—CBI can see out as far as the fourth and possibly the fifth peak. In the last week of May 2002, the CBI team released its first results. And once again, inflation passed a crucial test. "The beauty of CBI," said Anthony Readhead, a Caltech astrophysicist who heads the project, at the time, "is that we can chuck away the first peak entirely, and concentrate on the others. And looking just at those other peaks, we come up with the same numbers for the cosmological parameters that BOOMERANG and DASI and MAXIMA got from the first peak alone." It would be possible, though unlikely, he said, that the first peak didn't really point to a flat universe at all, that it just happened to be in the right position purely by coincidence. "It would be totally bizarre, however, if the parameters you deduce from the first peak and those that you deduce from a different scale entirely were both misleading. Nature would have to be doubly perverse to mislead you to the same conclusion twice."

So the universe really is flat, and really does have as much as 5 percent of omega in baryons and 30 percent dark matter—but again, "more or less" is an important modifier. "We can say we live in a flat universe to within 5 percent," said Readhead shortly after the CBI results came out, "and with better analysis of our data we can get that down to 3 percent or even lower. But MAP will look at the whole sky, and will do it as well as it can be done." MAP could still find some small significant deviation from flatness, he said, or from a scale-invariant spectrum of initial fluctuations (not realizing that it already had), and that could be terribly important. "Small deviations from the simple inflation-based model," he said, "could be where the clues to the physics lie. If MAP simply confirms what we have now, it will still have much tighter error bars, and that alone will make it a profound, significant experiment. As an observer, though, I've always hoped that we would find some unexpected things."

In fact, CBI did find something unexpected, a result that was already apparent in the team's preliminary data release circulated to the astronomical community a few months earlier, and which hadn't gone away. The amplitude of the third, fourth, and fifth peaks—that is to say, the height of the highs and the depth of the lows—was significantly higher than theory said it should be. "I can think of a few explanations," said David Spergel, who had been asked by Readhead to act as an independent commentator at the press conference. (Spergel had to pass because the Princeton astrophysics faculty had to discuss a graduate student who was going to be asked to leave; as the department's graduate representative, he couldn't be absent. They got Alan Guth instead.)

"If it's due to something primordial in the early universe," he says, "it could be an admixture of cosmic strings"—that is, strings of pure energy, light-years long, that some theorists have suggested could have been created as the universe cooled. That, he thought, wasn't too likely. A much more probable explanation would be a larger-than-expected number of galaxy clusters at a redshift of one or two, when the universe was half or less of its present age. Primordial microwave photons that scattered off electrons in these clusters would create cold spots in the CMB, contaminating the signal at the smallest angular scales.

"But I think it'll turn out to be something else," Spergel said. "I think the universe was reionized a lot earlier than everyone assumes." Reionization, recall, is the process that ripped some of the universe's hydrogen atoms apart again after they'd first formed at the same time the CMB radiation was emitted. MAP's polarization detectors were designed to pinpoint just when that happened. And although the analysis of the data had been under-way only for a couple of months, Spergel was ready to declare by June 2002 that "we've already found it. We see evidence of reionization at a redshift of 20 or so"—that is, when the universe was about 5 percent of its present age. "I'm going to NASA headquarters week after next," he said just before the Princeton reunions got under way, "to argue that MAP should be funded for an extra two years, beyond the original two. One great argu-ment would be that we've found the reionization and need the extra time to study it. But since we're not done with the analysis, I can't talk about it, even to NASA."

The most direct significance of such a discovery would be that the first generation of stars turned on about 50 million years after the Big Bang (by contrast, humans appeared on Earth about 65 million years after the dinosaurs were wiped out). These first stars would have been huge, a million or even 10 million times as massive as the Sun. "By itself, this doesn't tell us anything about cosmology," said Spergel, "although it tells us a lot about the astrophysics of the early universe, which is an important and fascinating topic." To date, he explained, every cosmic microwave background experiment has been studying the universe at the time of recombination, a few hundred thou-sand years after the Big Bang. "But now, for the first time, we're using the CMB as a flashlight, illuminating things that happened between then and now."

Indirectly, however, the way reionization contaminates the CMB signal can say a great deal about the CMB itself. "If 10 or 15 percent of the microwave photons we see are not from a redshift of 1000 but rather at a redshift of 20," he said, "then that would explain why the peaks CBI sees are too high." It would also mean that every measurement of the CMB that was ever made is really lower than it appears. Which means that the surprisingly low amplitude MAP is seeing in the intensity of the CMB at large scales—the same effect COBE saw—is even lower than it appears. That could mean nothing much, or could mean that the physics of the very early universe is not what cosmologists have been assuming for the past twenty years. "These discrepancies," Spergel said, "are starting to get interesting."

Not interesting enough to make any public announcements, though. Spergel was pretty sure he's right about what MAP was seeing, but team policy and scientific integrity demanded that he be more than pretty sure. Wanting to avoid the fighting that had gone on over when to announce the COBE results, Chuck Bennett wasn't insisting on a specific release date. "I think December, give or take a month, is probably a pretty good guess as to when it'll happen," Bennett was still saying in early June. That's what he was telling headquarters, but he acknowledged that it could be earlier. "The principle is clear," he said. "We work through it until we're convinced we've got a right answer, and that it's all written down, and then there's no reason to delay." To be convinced, though, he and Wilkinson and the rest of the team had to make sure they'd thought of everything, and analyzed everything in every reasonable way.

But so far, they weren't convinced, and they weren't talking. That didn't prevent their colleagues in the CMB community from talking, of course; it just prevented these outsiders from knowing what they were talking about. When an observation with such a huge potential payoff is in process, astronomers on the outside are about as resistant to spreading rumors as gossip columnists. Back in February, almost a year before the likely announcement date, Bennett had heard several already. One set of rumors coming out of a meeting on the West Coast said that

MAP would have nothing interesting to say. Another, out of Europe, insisted that MAP would have very interesting things to say. "I've been through this before," said Bennett. "It's inevitable." Sometimes, he said, people make the rumors up. Sometimes they get them by misreading something somebody says. Sometimes they start with somebody's guesswork which somehow gets garbled into a claim that it comes straight from a member of the team. "Basically I just ignore all that and do my work," said Bennett. "It's just part of the field, and it'll continue all through the year. We'll hear rumors that point in every direction. But it's way too early for us to know anything. And obviously nobody can know an answer that I don't know yet myself. It's kind of silly, but it's mildly entertaining."

Spergel even thought it might be fun to play a practical joke on the rumor-mongers. On the outside of his office door in Peyton Hall was posted a simulated map of the microwave sky, one of the many hypothetical maps generated with Gary Hinshaw's MAP simulation software. "It's not real," he said, "but statistically it's identical to what a real map might look like." When the actual map is complete, he could put the real thing up in place of the simulation, and he was convinced nobody would notice. "I love the idea that we could post the answer in plain sight weeks early and get away with it." If he were dumb enough to try it, though, they'd probably take away his MacArthur. During coffee hour at Peyton Hall, cosmology will frequently come up (no surprise, given Princeton's long involvement with the topic), and, said Spergel, "everyone kind of looks at me expectantly. I've been enjoying myself a bit. I say things like, 'So, if MAP were to find out that the Hubble Constant really is 40, what would you think of that?' I don't say this is what we've found, but I throw out numbers at the extremes of what people

The MAP Science Working Group takes a break during its July 2002 meeting at Jadwin Hall on the Princeton University campus. *Back row, left to right:* Eiichiro Komastu, Chuck Bennett, Lyman Page, Ed Wollack, Norm Jarosik, Al Kogut, Mike Nolta, Mark Halpern, Gary Hinshaw. *Front row, left to right:* David Wilkinson, David Spergel, Chris Barnes, Hiranya Peiris. *Missing:* Licia Verde, Ned Wright.

are expecting, and see what happens." At some point, in fact, probably in the fall, he was planning to post a challenge on the MAP website. Astronomers would get a chance to bet on the six cosmological parameters that MAP will nail down. "The closest answer will win something nontrivial," he said, "like a $200 bottle of wine." With that sort of prize, Spergel figures he'll get some pretty thoughtful estimates. "It's a good way to get a measure of what the community's really thinking," he said.

At least some people were convinced that MAP will find nothing interesting at all. People like Robert Kirshner at Harvard and Michael Turner in Chicago continued to think that MAP would simply refine the numbers cosmologists already had from BOOMERANG and CBI and the supernova experiments and other sources. In Britain, a theorist named John Peacock, at the University of Edinburgh, was reportedly telling his colleagues that cosmology was already solved; there was no need to wait for the MAP results. "I'm happy to see him saying this," Spergel said. Why? "Let's just say it will be useful to refer to in future talks."

But a time would inevitably come—also in the fall, presumably—when the team would know the full answer, not just the hints that had shown up so far. At that point, it might not be so much fun to hang out with other astronomers. "I'm a little nervous about this meeting I'm going to in early November, in California," said Spergel. "I'm scheduled to give an overview of the current status of cosmology, and by then we'll almost certainly be dotting the T's and crossing the I's in the wording of the paper. And everyone in the audience will know that I know. It's going to be a very awkward time." In fact, he was considering not going. "I don't know how useful it is for me to go to a conference and say, 'I can't talk.' "

By late June, it was already getting a bit awkward dealing with NASA headquarters. Spergel, Chuck Bennett, and Gary Hinshaw had gone in to make their case for an extra two years of operation. "They really wanted to see some data," said Spergel, "partly out of curiosity, but also because they'd like to be reassured that the satellite really is working." The NASA administrators were willing to take the team's word for it, given Chuck Bennett's reputation, but they would have been happier with actual measurements. Spergel now had another worry as well. "I got a call from our system administrator asking about my attempting to log in on Saturday night. I wasn't working on Saturday night." Someone had hacked into the team's server. At first everyone was worried that it was targeted at MAP, that someone was trying to alter or steal the data. That probably wasn't the case, since the first thing the hackers did when they got in was to try and leapfrog into other computers. "I doubt they could have found anything useful anyway," he said. "I have the data hidden in directories that have nothing to do with MAP. I'm also bad at labeling things—the Goddard guys complain that they can't understand the figures I send them. The real problem is that I haven't had access to the computers for a week now, and I'm falling behind schedule."

Officially, then, Spergel knew nothing. Unofficially, he predicted that a lot of cosmologists could be in for a shock when January rolls around. "There's a certain conservatism in cosmology." Anthony Readhead's open-mindedness notwithstanding, he said, "Just about everyone thinks we have the right model, so everyone expects MAP to be uninteresting." Yes, the BOOMERANG and MAXIMA results didn't match each other precisely, but everyone assumed the Standard Model, incorporating cold dark matter, inflation, and a large dose of dark energy, would fall into place. "As of now," said Spergel, "it looks like that won't happen. People keep overlooking the fact that we have two tooth fairies, in the form of CDM and lambda [i.e., dark energy] that we don't understand at all. And so it shouldn't be all that astonishing that we're finding that the Standard Model doesn't fit after all."

In fact, he said, that isn't necessarily all that surprising, given that the key element of the Standard Model—inflation—came out of the Grand Unified Theories (GUTs) of the 1970s. The motivation for these theories was to explain how the forces of nature (the strong and weak nuclear forces, and electromagnetism and gravity) are really just different aspects of a single, primordial force that dominated the universe in the very beginning. But reconciling gravity with the other three turned out to be extremely tough, and during the 1980s and 1990s physicists began to consider an entirely different class of theories. Known originally as string theories, they're based on the idea that the building blocks of matter and energy aren't point particles, but rather tiny loops of stuff that vibrate at different frequencies. These "superstrings," moreover, exist in a space with ten or more dimensions; we don't see most of them because they're only discernible at very small scales (the often-used analogy is that a garden hose has thickness when you get close, but from a distance looks like it's a single, thin line).

As superstring theory has evolved into a more complicated construct known as "M-theory," physicists have postulated that our entire three-dimensional universe could be floating within a higher-dimensional space, something like a sheet of paper floating in the wind. And if M-theory really is a better way to explain the structure of the universe than GUTs were, maybe it could also come up with a better explanation for how the universe started out. One serious problem with the Standard Model has always been that it implies a singularity—a point, right at the very beginning, when the density and temperature were infinite, and when time began. Physics has no way to describe such a situation; so while the universe is the effect, nobody had any way of addressing the cause.

This really bugged Spergel's Princeton colleague Paul Steinhardt. Steinhardt is considered one of the fathers of inflation; nevertheless, he wondered whether there might be a way to create the universe we see without resort to such an unsatisfying nonexplanation—and without inflation. Working with Neil Turok, Spergel's former collaborator, he came up with an idea that comes out of M-theory, not old-fashioned GUTs. In this formulation, the universe is analogous to a two-dimensional sheet of rubber that floats along in a three-dimensional space (though in our case, it's three dimensions floating in four). There's another sheet, floating parallel to it. Our sheet—our universe—is empty; the density of matter is zero. Now the two sheets (or "branes," short for "membranes") collide. The resulting release of energy drives the temperature of our universe to perhaps a 100 billion billion degrees, and starts our brane stretching—still flat, but stretching. As it stretches, it cools. Matter congeals and forms into galaxies. Eventually, the cooling of the vacuum dumps extra energy into the system, just like inflation but a hundred orders of magnitude less intense. "That," says Steinhardt, "accounts for the so-called dark energy that we observe accelerating the expansion of the universe."

As the dark energy forces our expanding sheet to expand ever faster, matter is eventually diluted to the point where its density is zero. "That takes a few trillion years," says Steinhardt." But something else is also happening. The two branes have been exerting an attractive force against each other, and while they bounced apart after the initial collision, that force begins to draw them together again. At the same time, the dark energy begins to run out; our brane begins to collapse again. Quantum fluctuations, like those invoked in inflation, make different parts contract at different rates. And before the collapse gets very far, the two branes collide again. The temperature shoots up, the branes bounce apart, and our brane starts to stretch again. Radiation congeals in matter—with the fluctuations preserved. Presto: a new universe, with fluctuations that will show up in the microwave background. And then the whole thing repeats. Forever.

"With a Standard Model based on inflation," says Steinhardt, "you need two periods of acceleration, one at the beginning and one later on for the dark energy. So in a sense, the new model is more economical. And it avoids the rather difficult problem of explaining how there could be a beginning of time." Unlike Dicke's cyclic models of the 1960s, this one has cycles of equal length, stretching indefinitely into the past and into the future. The only downside of Turok and Steinhardt's model is that while it's simpler, more elegant, and less intellectually troubling than the Standard Model, its predictions are no different for any observations you can make today—even with MAP. It will take another generation of CMB detectors, at least, to tease out the slight differences between the two theories. But it's intriguing enough that cosmologists are paying it a lot of attention. Even a hard-nosed experimentalist like Dave Wilkinson said: "I have no evidence that it's right . . . but it just smells right to me."

But even though MAP can't say anything about the Turok-Steinhardt model, it was pretty clear by late summer that the Standard Model could be in some trouble. Evidence of early reionization and of deviation from scale invariance was showing up in all of the individual maps constructed from measurements at different wavelengths. You can compute the power spectrum of fluctuations from the individual maps one by one, or by combining them in different ways—fifty-five of them, by Spergel's count. "Ideally," he said, "you should get the same answer from all fifty-five." Practically, it wasn't happening yet. This was to be expected because the experimentalists were still teasing out sources of extraneous noise. On Pass Zero, at processing the data, they made the assumption that overall noise levels were constant from one month to the next. "But we know the Earth's orbit is elliptical, not circular. In December, you're closer to the Sun, so MAP is 1 percent warmer, and a warmer detector is noisier." On subsequent passes at the data, subtleties like this were factored in, but subtler subtleties would remain until the final analysis. Lyman Page was still looking at the observations of Jupiter to model the shape of the beam with exquisite accuracy, for example; when he was satisfied, the data would be reanalyzed in light of what he found.

If the Standard Model fit after all, MAP would be able to give terrifically precise measures of the Hubble constant, the density of baryons in the early universe, the ratio of normal to dark matter (whatever it is), the curvature of the universe, the amount of dark energy (whatever it is). The press conference in January would present and celebrate these numbers. Cosmology would be, if not finished, then at least in its final chapter. If the Standard Model didn't fit—if standard inflation plus cold dark matter plus lambda were inconsistent with the maps no matter how you adjusted these parameters, then cosmologists would need a new model—a new context for understanding the universe. "I've been talking to Chuck about how we should handle this," Spergel said. "We clearly need to be ready to suggest some alternate models." Offhand, he can think of four. One would be to presume there's an extra ingredient to the early universe that people haven't been factoring in—a much bigger dose of massive neutrinos than anyone had allowed for, or a bath of gravity waves from some unexpected process.

A second was the idea Spergel had floated in June: an extra inflationary field. Standard inflation would leave its imprint on spacetime in the form of scale-invariant density variations that were also adiabatic—meaning that both ordinary baryonic matter and weakly interacting dark matter were dense in the same places and sparse in the same places. Their densities would evolve differently, since baryonic matter feels the pressure of photons and dark matter does not. But they'd start out the same. If there was a second inflationary field, though, it might not have been adiabatic. It could instead have exhibited a property known as isocurvature, in which dark matter and baryonic matter would complement each other, one dense where the other was sparse and vice versa. A second field of this kind, said Spergel, could account for the discrepancies MAP seems to be turning up.

If that were true, however, the imprint of isocurvature fluctuations could also spell trouble for the astrophysicists who are trying to detect the particles they're convinced make up the dark matter. If these particles really are weakly interacting—if they participate in the weak nuclear interaction—their independent imprint should have been wiped out by the time of recombination, as dark-matter particles would meet and annihilate one another and re-form from the resulting energy. If the imprint is still there, this suggests that they may not experience the weak force after all. Since they also don't feel electromagnetism or the strong nuclear force, that leaves only gravity. And since the minuscule gravity of a single particle is undetectable by any known or contemplated measuring device, that means the dark matter that makes up 90 percent of the mass of the universe might never be detected at all.

Spergel's third idea was that even a single inflationary field might be more complex than everyone now thinks, leading to a more complex pattern of primordial fluctuations. And finally, and most tantalizingly, he hadn't given up on the idea that the universe might be finite. "We've finally got all our code ready to go," he said at the beginning of August 2002. "But the truth is, I'm afraid to push the button." Why? Because if the universe isn't small after all, he'd be a little disappointed, even though this was the result he expected. And if it is, the world of science will be shaken as deeply as it was when Copernicus showed that the Earth isn't the center of the universe. Spergel knew that as the agent of that change, his life would never be the same again. "I'm taking the family to Hawaii for a vacation in a couple of weeks," he said. "So what I'm thinking of doing is setting everything up, pushing the button, and then leaving. I can check in from an Internet cafe in Hawaii and see how it all turned out."

CHAPTER $\boxed{12}$

The Answer

In the end, David Spergel decided not to log in from Hawaii. He wasn't convinced, after all, that the software was completely reliable. Moreover, he realized that the answer might ruin his vacation. Learning that the universe is not finite wouldn't be such a problem; he figured this was the most likely outcome, anyway, and he could live with it. But if the computer reported back that the tell-tale signature of a finite universe had shown up in the CMB, he'd spend the rest of the time torn between excitement and the fear that there was still some effect he hadn't properly accounted for. Palm trees, glistening sand, tropical breezes would all fade into the background. He couldn't in good conscience do that to his family.

Then, soon after he returned to Princeton, all thoughts of the finite universe and the CMB and MAP were driven from his mind, and from the thoughts of Chuck Bennett and Lyman Page and everyone else on the team. At the end of August, Dave Wilkinson entered the hospital, as he had dozens of times during his nearly twenty-year battle with lymphoma. On all of his previous visits, he'd been treated and released, spent a few weeks getting back his strength, and jumped back into work. This time was different. On September 5, 2002, he passed away. Two facts about Wilkinson's death are enough to sum up his life. First, he postponed checking into the hospital for several hours, against his doctor's advice. *Discover* magazine was preparing a feature story about the project to search for alien laser-light signals, and a photographer was coming out that day. Ill as he felt, Wilkinson thought he'd be letting the team down if he didn't show up for the pictures. (In the end, *Discover* ran a picture of Wilkinson alone, not with the group. He would have been indignant at being singled out.)

The second fact was that Wilkinson's memorial service had no fewer than twenty-one speakers who stood up before the crowd that packed the Gothic vastness of Princeton University Chapel and testified to his kindness, his generosity, his importance to their personal and professional lives—and finally, to his enormous influence on the science of cosmology. "The one good thing," says Chuck Bennett, "is that Dave did live long enough to see a version of our sky maps, even if it wasn't the final version. He was delighted, had a real sense of accomplishment." Months before they agreed on anything else they would say at their public announcement, the MAP team voted unanimously to ask NASA to rename the satellite. The agency was pleased to go along. The formal name would henceforth be WMAP—the Wilkinson Microwave Anisotropy Probe. For someone like Dave Wilkinson, it was perhaps even more appropriate than a Nobel Prize.

The team could be distracted by grief for only so long, however. They had data to process, and reprocess, and test over and over again. Gary Hinshaw had constructed thousands of simulated maps of the microwave sky, each one based on a subtly different mix of parameters—a little more dark matter here, a bit more ordinary matter there, an expansion rate that was slightly faster or slightly slower, a universe departing in one direction or other from perfect geometric flatness. The actual map would have to be tested against each of these—not directly, since even a perfect fit wouldn't be identical to the real map, except in a statistical sense. But before this could happen, the actual map would have to be as free of imperfections as possible. Did the team account for every conceivable source of contaminating radiation? Would they miss something and face deep embarrassment when someone else reanalyzed their data?

"We were all concerned about this stuff," says Spergel. "But Chuck stood out. I've never seen anyone worry like he did. The pressure was enormous. At the last science team meeting, Lyman said, 'It's amazing. Not only did the thing work, but we're still talking to each other.' It's like having a small child," Spergel continued—with a fair amount of authority, given that Joshua, his third child, was nearly a year old. "You know, it's a very special time. But you're so busy and so tired, it's hard to really appreciate it until afterward." In preparation for that last meeting, Spergel had gone to Small World Coffee (a locally owned coffee bar that discriminating Princetonians patronize in preference to the cookie-cutter Starbucks around the corner) and bought T-shirts for the team. "They read: Sleep is for the Weak."

The formal announcement date, meanwhile, began to slip. Originally, Chuck Bennett had hoped to present MAP's results at NASA headquarters on January 2, 2003. But the analysis took a bit longer than he'd thought. It would have to be pushed back a week or so—except that with NASA, you can't just push things back. The agency has all sorts of projects going, and it doesn't like to put out too many announcements in too short a time. It takes away from the impact of each. So the MAP announcement was shoved back to the end of January—except that January 28 was the date for the President's State of the Union address, and given the approach of a possible war with Iraq, that could be too much of a distraction. February 6, a Thursday, looked good. Nothing else was going on. Unless the war had actually begun by then (which nobody expected), MAP's results should make plenty of news.

No one expected a war—but no one expected that the Space Shuttle *Columbia* would explode over Texas just five days before the planned press conference, either. NASA suddenly had plenty of other things on its mind. There was no way the press conference could come off on time. But while the MAP team was itching to get its results out, they knew that this delay, perhaps unlike the others, was absolutely appropriate. "Listen," said Lyman Page later that evening, "the importance of what MAP has done is going to play out over years. Whether the announcement comes this week or even this month won't make any difference."

That was all the more certain because the team wouldn't be saying anything shocking. With its fine resolution, all-sky coverage, and spectacularly exhaustive error analysis, the satellite might have found that earlier, ground-based experiments had been badly fooled. It might have found shocking evidence that inflation never happened, that some bizarre new species of darkness was required, along with dark matter and dark energy, to explain the structure of the cosmos. It might have found that the universe is finite.

But it didn't. "I have finite-universe models," says Spergel, "that are consistent with what we see, but only if the finite universe is just a bit bigger than the visible universe." If that were the case, you'd have to look just a bit farther than it's possible to look in order to see nearby galaxies a second time, far away. Not only that: the existing Standard Model of cosmology looked like it was basically right, after all. What MAP had done was to put precise numbers in places where there was only a range of numbers in the past. "In a way," says Spergel, "it's slightly disappointing. We can fit just about all the existing data from all sorts of different observations in many different wavelengths into a single, consistent picture. We really do live in a flat universe, with dark matter and dark energy. It would have been more fun to find a big surprise."

Instead, he suggested in early February, just before the press conference had been postponed, that the headline in the *New York Times* would read: "Five Numbers Explain the Universe." MAP had determined to better than 2 percent accuracy that the age of the universe is not between 12 and 15 billion years, as cosmologists had been saying for the past half-decade, but 13.7 billion years. Ordinary atoms account for just 4.4 percent of the universe, dark matter makes up another 23 percent, and the rest is dark energy. The Hubble constant, which measures the current expansion rate, is 71 kilometers per second greater for every megaparsec (that is, every 3200 light-years or so) of distance you look farther from Earth. And the time when the first stars turned on is somewhere around 200 million years after the Big Bang— a bit later than Spergel's ballpark estimate the previous summer. Even so, "That last number is pretty surprising," says Spergel. "Most people, though not all, have argued that it took place closer to a billion years after the Big Bang."

Just as Spergel had suspected back in July, this early lighting up of stars, revealed by patterns of polarization in the microwave background radiation, created an extra burst of energy that smoothed out all of the peaks of temperature, making them a little lower than they'd started out, and elevating the lows; the result was a pattern of fluctuations about 30 percent more violent than they ultimately appeared (this was the opposite of what Spergel seemed to be saying the previous summer, but that was due to confusion on the part of his audience). That's why, when people like John Peacock said all the answers were already known, Spergel said they were in for a surprise. "Things weren't radically off," Spergel admits, "but they were systematically off."

This kind of revision makes astrophysicists sit up and take notice. For the general public, though, the MAP result involved no visions of God, no Holy Grails. Neither Spergel nor Bennett is likely to get a call from a high-powered literary agent. But the public view of science doesn't always have a lot to do with the real thing. It's true that scientists are out to understand how the world works—but such understanding doesn't come in a thunderclap. A specific result, no matter how surprising or fundamental, is really just a starting point for the real work of science.

The science writer John Horgan argued in his book *The End of Science* that we've already made all the big, fundamental discoveries, or most of them, anyway. But those discoveries—that evolution proceeds by natural selection, that DNA is the molecule of heredity, that the universe began in a Big Bang, that the chemical elements are arranged in a periodic table, that atoms exist and operate in certain ways—all of these provide frameworks within which scientists can explore the crucial details. Just to name a recent example, molecular biologists announced three years ago that they'd sequenced the human genome. It was a historic milestone—but it told scientists nothing about precisely where the genes are located along those endless stretches of DNA, what protein each gene governs, what these proteins do, or what the biochemical fallout from interactions among proteins might lead to. Horgan had a catchy title, but science isn't just the making of great discoveries; it's understanding what they imply.

And so it will be with MAP. "Journalists will probably say it's a 'dog bites man' story," Max Tegmark commented just before the announcement (he claimed to have heard solid rumors from reliable sources, whom he would not name). "Things are pretty much as we thought. But to a scientist, there's an enormous difference between believing something is true and knowing it's true. The Standard Model was extremely wobbly before, very floppy. MAP has tightened the bolts." Now, armed with precise numbers where they once had ballpark figures, cosmologists will be able to start solving some of the key unanswered questions of cosmology. "It's very much like the Standard Model of particle physics has been for the last few years," says Spergel. "They have something like seventeen parameters, like the strength of forces and the masses of particles, that explain everything with high precision. But nobody knows why those things are what they are. Now we have a standard model of cosmology, and we can begin trying to understand the details."

"You have to understand," says Lyman Page, "that in one single set of observations we're getting information from the first 10–35 seconds of the universe, which is when the patterns in the polarization were laid down; from about three minutes after the beginning, in the proportions of dark and baryonic matter; from 380,000 years after the beginning; and from 100 million years after that, in our detection of reionization. That's incredible!"

Beyond that, he says, all of these numbers, which cosmologists knew to within 20 percent at best before MAP, are now known to within better than 2 percent. "It's like the universe just got smaller. We know it like we know our own backyard. In some sense, we know more about what's going on at the edge of the universe than we know about what's happening at the bottom of the ocean."

MAP's measurement of a flat universe, despite Spergel's initial speculations, does imply that inflation—or something very much like it—happened. "An extra inflationary field, it turns out, makes our fit a tiny bit better, but not enough better that we're confident about claiming it." In fact, MAP's measurement of cosmic flatness is much more definitive than anything before, not just in its precision, but because the fluctuations in polarized light over large swaths of the universe are statistically consistent with the fluctuations in temperature. "You could imagine some process much more recent than the era of decoupling and recombination that could have produced the temperature fluctuations," says Spergel, "but they couldn't have produced comparable fluctuations in temperature and polarization." (One more triumph of precision: recombination happened, says MAP, not at "about 300,000 years after the Big Bang," but at 380,000 years, give or take 10,000.)

This doesn't rule out the competing Steinhardt-Turok model that invokes colliding branes (which they gave the distinctly non-euphonious name "the Ekpyrotic Universe"). It differs from inflation so subtly that MAP can't distinguish between the two. But the pattern of fluctuations does rule out entire classes of inflation, including the simplest versions. "There's now going to be a cottage industry in nailing down inflation," Bennett says. The same sort of thing will happen with most of the remaining unanswered questions of astrophysics. What's dark matter actually made of? (Not neutrinos, anyway: MAP puts an upper limit of 0.2 electron volt on the neutrino's mass; its contribution to cosmic structure is negligible.) What is the dark energy? ("Quintessence," one variety of dark energy favored by some theorists, seems to be ruled out, and that's a good thing: unlike more conventional versions, quintessence can come and go, and even reverse direction. It could, in principle, cause the universe to start collapsing at high speed any time now.) Did inflation really happen, or was it indeed something like Steinhardt and Turok's Ekpyrosis? Despite what John Horgan claims, if all of this is not science, then it's hard to know what you'd call it.

Another important consequence will come out of MAP's discovery of reionization. "Not only does it contradict what a lot of people thought," says Spergel, "but at one stroke it makes the science case for building the James Webb Space Telescope." This is the observatory, originally called the Next Generation Space Telescope, that John Mather left the MAP team to work on. Astronomers had pushed to make this larger-than-Hubble telescope sensitive to infrared radiation, not visible light, on the theory that all the action of star and galaxy formation occurred very early in the lifetime of the universe, during a period that's come to be known as the Dark Ages. That being the case, any light from that formative era would now be redshifted so strongly that it would peak in infrared wavelengths. MAP says they got it right. "I think we've probably defined the agenda for extragalactic astronomy for the next decade," says Spergel.

The MAP team has also defined its own agenda. The project is hardly over; with nearly three more years of data taking and processing, most team members will be spending significant parts of their time on the satellite for which they've lived and breathed for the better part of the last decade. It's not just busy-work, either. The continuing temperature measurements will re-fine MAP's precision on cosmological parameters, giving theo-rists an even more solid platform from which to build models. The polarization measurements, which are currently pretty crude, will improve dramatically. They may even rescue Sper-gel's most radical idea. "If the universe really is finite and a bit larger than the visible universe," he says, "it turns out we could still detect it." That's because in a finite universe the microwave background would "ring" like a bell, with characteristic fre-quencies. And this ringing could show up in the pattern of polar-ization. "The signal-to-noise ratio should improve fourfold in four years," he says. "I thought that we were out of luck."

Whether or not the absence of fluctuation on the largest scales means that the universe is finite, it's at the very least puzzling. "If it's significant," says Bennett, who tends to be less outspoken than Spergel, "it's very important, if you know what I mean." Page agrees. "It's not a huge effect, but it's just unsettlingly weird. It's not the sort of thing that overthrows a paradigm, but it does say that there's more to it than you're thinking right now."

Spergel's ringing bell may provide the solution, but it may instead come from Page's next big project. He and a group of collaborators are on the verge of getting funded to build his six-meter microwave telescope in the thin air of the Chilean Andes. That instrument will probe fluctuations at very small scales—and the combination of small-scale and large-scale measure-ments could straighten everything out. Inflation theories tend to predict a flat spectrum of primordial fluctuations overall, but funny things tend to happen at the edges. The simplest models suggest an excess at very large scales, which is why MAP and

COBE's deficits are so odd. Once the small-scale fluctuations are well understood, the picture might make more sense.

Page and Bennett and several others are also talking about the next satellite project. Although the European Planck satellite should refine MAP's measurements even further, the action in cosmological research is shifting away from microwaves and toward polarization. The polarization detected by MAP (and first glimpsed a few months earlier by DASI, at the South Pole) comes from light bouncing off electrons during the early universe; it indirectly probes fluctuations that started out long before the universe was one second old.

In order to test inflation definitively, however, you need a direct line to that earliest of moments. It turns out there is one. Inflation would not only have set up density fluctuations in the newborn universe; it would also have generated gravity waves, which would still be vibrating through the universe. A collision between two branes, as in the Ekpyrotic Universe model, would not have generated these waves, and finding or ruling them out is the best way to figure out which is more plausible. The waves themselves are extraordinarily hard to detect. (Two enormous detectors now being tested in Louisiana and in the state of Washington may not even be sensitive enough to pick up the far more violent waves generated by colliding black holes.) But they would have generated a distinct, though vanishingly faint, pattern in the polarization of the CMB. That's what Bennett, in particular, is thinking about next. "Building a satellite to detect that signal will be a lot more difficult than building MAP. But I think it's possible, and several of us are starting to toss around ideas of how we might do it."

Spergel expects to be involved at some level with both of these projects, but they probably won't be his major focus. "I'm being asked to serve on all sorts of committees, too, and I'm working with a lot of students. What's clear," he continues, "is that I won't be doing any small projects anymore. I did that for the first half of my career. Now I'm in the second half, and that's over. On a five- to ten-year timescale," he says, "I'll mostly be involved with looking for planets. I had an idea of how to make our new light-suppressing telescope even better, but we're not submitting the paper until after MAP comes out. I haven't worked on anything but MAP for the past six months—but it might look otherwise." A few years ago, Princeton's astronomy department was so focused on cosmology and so clueless about extrasolar planets that a world-renowned visiting expert on the latter was treated to a discussion on the former. Now Spergel and several colleagues are plotting a bid to have NASA base an astrobiology institute on campus. "You have to admit," he says, "that it would be very cool fifteen years from now to announce: 'We just saw the first earthlike planet orbiting another star.'"

Epilogue

By the summer of 2004, things had calmed down considerably for the WMAP team. For the first several months after the first results were announced, Spergel, Bennett, Page, and the rest were spending virtually all of their time furiously redoing calculations, refining their understanding of the noise contaminating the signals they were seeing, and driving the uncertainties in the measurements even lower than they already were.

At the same time, other cosmologists dived into the ocean of data and began doing their own analyses. "At first," says Bennett, "there was a natural tendency to poke at our results, to try and find out what we might have gotten wrong. Generally, the people who were quickest to do that didn't have time to do so in detail, so while there were some disagreements early on, as time went on, pretty much everyone agreed that our results were basically correct.

"By now," he says, "we've counted well over a thousand papers that reference our original papers, which makes them some of the most highly referenced papers in physics and space science." Being highly referenced isn't something that the general public is likely to care much about, but among scientists it's the most gratifying vote of confidence possible. It means your colleagues broadly consider your work to be important.

For the WMAP team, that's what matters. The original, extraordinarily ambitious idea was to make the most comprehensive and definitive measurements of the CMB to date and to lay to rest any significant uncertainties in the basic parameters of cosmology. WMAP has done that with astonishing success. It would have made for better front-page headlines if the satellite had found something truly revolutionary, but that wasn't ever the expectation. Nor, despite the transitory glory that goes with such a discovery, was it high on anyone's list of hoped-for outcomes. It's difficult for non-scientists to really grasp the fact that discovering the truth about the natural world is, in most cases, the real goal of science.

The satellite's resounding success doesn't mean, however, that all of WMAP's results have been fully explained. There's still the riddle of the lack of CMB fluctuations on the very largest scales. A few months after the original results were released, an independent team of astronomers and mathematicians published a paper in *Nature* declaring, based on WMAP, that the universe is indeed small, with a topography analogous to that of a dodecahedron—a soccer ball.

But an exhaustive analysis by Neil Cornish, of Montana State University, in collaboration with Spergel and others, eventually ruled out that and many other possibilities. "We had the dodecahedron, the trombone horn, the torus," says Bennett. "But they all turned out to be incorrect. It's still not impossible that the universe has a compact topology. But it's looking increasingly unlikely."

Now, nearly two years after the first announcement, most of the team is winding down its work on WMAP and is thinking about, or actively working on, other projects. Lyman Page, for example, is deeply into the planning of the Atacama Cosmology Telescope (ACT), a 6-meter-diameter radio telescope to be built in the high desert of the Chilean Andes at an altitude of about 16,000 meters. Within a few years, Page and his collaborators will use ACT to reliably measure the CMB to much smaller angular scales than anyone has done before.

Ned Wright had already pulled back from his work on WMAP to serve as principal investigator on a different satellite, the Wide-Field Infrared Survey Explorer, which NASA had approved before WMAP was launched.

Al Kogut is still spending about a third of his time on WMAP, focusing mostly on the data on polarization. The rest of his time is split between two new balloon-borne instruments, one to measure the polarization of CMB microwaves more precisely in order to search for the signature of gravity waves, and the other to measure the heating of the universe caused by the ignition of the first stars.

Like Kogut, Ed Wollack is devoting about 30 percent of his time to WMAP, but he's also designing microwave detectors for new CMB missions. Also like Kogut, his new interest is in finding ways to measure polarization.

Polarization is on Gary Hinshaw's mind as well. He's still working mostly on WMAP, trying to understand what it can reveal about CMB polarization. "I am also," he wrote in an email in September 2000, "leading a study team of about 25 scientists (including most of the WMAPpers) who are interested in building a mission to map the CMB polarization with much higher sensitivity to study the physics of inflation."

Bennett, too, is still spending much more than half his time on WMAP, but he is beginning to think about other projects as well. He is, for example, one of the twenty-five scientists on Hinshaw's study team. He is also chairing his own science team that is trying to figure out what sort of space-based missions could best answer the question of whether dark energy is changing in strength over time. "I think dark energy is *the* big problem in astrophysics now. What is it, and what was its influence in the early universe?"

Spergel, meanwhile, is on sabbatical at the Cerro Tololo Inter-American Observatory in Chile. Predictably, his intellectually restless spirit is moving in several different directions at once. In an email shortly after he arrived there, he wrote,

> I have spent some of my time here preparing for the next steps with WMAP and beyond. With 8 years of WMAP data, we should have the sensitivity to detect gravity waves if we can remove the foreground emission from dust in our Galaxy. My plan is to use observations of bright stars in our own Galaxy to trace out the dust distribution. I have been talking to several people here about carrying out an optical survey in the South. Much of the work may be done with 16-inch telescopes! We also hope to use the McDonald Observatory in Texas.
>
> I am also spending some time preparing for the ACT experiment and will visit the ACT site (mostly to see the desert) next week.
>
> I am planning on starting some new project later on during the sabbatical.
>
> Fred Adams and I are writing a speculative paper on panspermia
> . . .
>
> I want to learn more about gravity waves . . .
>
> I also want to learn more about planetary atmospheres . . .
>
> I don't know where these projects will go next.

As for WMAP itself, the satellite is still chugging along happily, a million miles from Earth, scanning the CMB over and over to tighten up the temperature and polarization maps still further. At any point in its long history, from the first gleams in Bennett's and Wilkinson's eyes to the competition with other teams to the construction and launch and deployment in space, a hundred things could have gone wrong, and the mission would have been over before it measured anything at all. Faulty screws in the detectors, faulty power converters in the satellite hardware, a misfiring rocket . . . just about anything could have torpedoed the mission.

But nothing did. Against formidable odds, WMAP worked just about perfectly. "Originally," says Bennett, "it was supposed to be a two-year mission. When it came up for review, we asked for another four. And most recently, we've asked for funding to extend it to eight years, total. NASA has tentatively agreed." Chuck Bennett is not a publicly demonstrative man, and WMAP is, after all, just a piece of hardware. But his personal pride in the little satellite, the child of his intellect and grueling work, can't help but show a little. Nor can a small twinge of sadness as he adds, "we won't ask for more than that."

Glossary

acoustic peak — A point on the spectrum of CMB fluctuations where pressure waves overlap to reinforce each other, creating a local maximum.

adiabatic — An initial pattern of fluctuations where matter of all types is clustered at the same places.

aether — A hypothetical medium that pervades the universe, and through which light propagates as waves propagate through water. Its existence was disproven by the Michelson-Morley experiment in 1887.

anisotropy — Of different strength in different directions. The distribution of stars in the sky is anisotropic, since most are concentrated in the Milky Way.

baryon — One of the particles—e.g., protons or neutrons—that serve as the building blocks of the chemical elements.

Big Bang — The event that marks the birth of the universe, approximately 14 billion years ago.

blackbody curve — A plot that describes the mix of light wavelengths coming from an object (including the largest of all objects, the universe) in thermal equilibrium. The shape of the curve depends only on the object's temperature.

blueshift — A compressing of electromagnetic waves caused either by an object's rapid motion toward the observer or by the shrinking of spacetime between observer and object.

bolometer — A device that converts electromagnetic radiation into heat with high fidelity, allowing the energy to be measured with great precision.

CMB — The Cosmic Microwave Background.

COBE — The Cosmic Background Explorer satellite. In 1990, it confirmed that the cosmic microwave background conforms perfectly to a blackbody curve, and in 1992, it found definitive evidence that the radiation reflected a pattern of primordial density fluctuations.

Cosmic Microwave Background — Electromagnetic radiation emitted at the time of decoupling, now redshifted down to a temperature of approximately 2.7°C.

cosmological term or cosmological constant — An extra term introduced into the equations of general relativity by Albert Einstein to account for the fact, which seemed evident at the time, that the universe is not expanding or contracting.

critical density — The amount of matter per unit volume that would generate just enough gravity to counteract cosmic expansion without causing contraction.

dark matter — A yet-unidentified form of matter whose mass is at least ten times greater than that of the visible stars and galaxies.

decoupling — The event, about 300,000 years after the Big Bang, where electromagnetic radiation was finally able to travel through the universe, unimpeded by collisions with charged particles. It happened because charged electrons and atomic nuclei merged to form neutral atoms (*see* recombination).

dipole — The illusion that the CMB is hotter in one half of the sky than in the other, caused by Earth's motion through the universe.

DMR — One of COBE's three main instruments, the Differential Microwave Radiometer measured fluctuations in the CMB.

Ekpyrotic Universe — An alternative to inflation, it posits that the Big Bang was actually a collision between three-dimensional "branes," that the universe experiences recurring Big Bangs. Unlike the Standard Model, this theory does away with the need for a beginning of time and space.

electromagnetic spectrum — The range of varieties of light, including, from lowest to highest frequencies, radio waves, microwaves, infrared light, visible light, ultraviolet light, X-rays, gamma rays.

HEMT — High Electron Mobility Transistor, a solid-state amplifier at the heart of the MAP satellite's radiometers.

flatness problem — The uncomfortable coincidence that of all possible values of omega, the matter in our universe is within an order of magnitude of being precisely 1.

horizon problem — The uncomfortable fact that opposite sides of the universe are at precisely the same temperature, even though under ordinary circumstances they couldn't ever have exchanged heat.

inflation — The proposition that in its very earliest moments, the universe expanded at a rate many orders of magnitude faster than the speed of light; solves both the horizon and the flatness problems.

inflaton — The energy field that drove inflation.

isocurvature — An initial pattern of fluctuations in which matter is distributed smoothly, but different types of particles cluster in different places.

isotropy — Smoothness; identical in all directions.

L2 — One of five Lagrangian points where the gravity of Earth and Sun are balanced, and where objects tend to remain without much help. If you extend the imaginary Earth-Sun line about a million miles out beyond Earth, you're at L2.

monopole — A hypothetical particle, analogous to the electron or the positron, that bears just a north or just a south magnetic charge.

omega — Also known as the density parameter, it is the ratio of the universe's actual density to the critical density. At critical density, omega equals 1.

photon — The smallest unit of electromagnetic energy, and thus of light; conventionally thought of as a particle.

plasma — A gas so hot that electrons are stripped from atomic nuclei, forming a cloud of charged particles. Until about 300,000 years after the Big Bang, the universe was a cloud of plasma.

polarization — The orientation of light waves into planes of oscillation when they are scattered.

Primeval Atom — The term Georges Lemaître used in the 1920s to describe the dense knot of matter from which the expanding universe leapt.

radiometer — An electronic device that measures the intensity of microwaves with high precision.

recombination — The event, about 300,000 years after the Big Bang, when the universe had cooled enough that electrons and atomic nuclei could combine to form neutral atoms, permitting photons to be decoupled from matter and travel unimpeded (see decoupling).

redshift — A stretching of electromagnetic waves caused either by an object's rapid motion away from the observer or by the stretching of space-time between observer and object.

reionization — A period, some time after recombination and decoupling, when ultraviolet light from hot young stars ripped apart hydrogen atoms.

scale invariance — The phenomenon of fluctuations that are of about equal intensity at all angular scales.

spacetime curvature — The notion, introduced by Einstein, that spacetime has its own geometry, dictated by the density of matter within it.

spectrum of fluctuations — The mix of different-sized ripples in the density of the early universe, and thus in the temperature of the CMB.

spiral nebula — Pinwheel-shaped objects that some astronomers thought were clouds of hot gas within the Milky Way. Edwin Hubble proved in the 1920s that they were actually full-fledged galaxies like our own.

Standard Model — The conventional cosmological wisdom. It states that the universe started with a hot Big Bang, that inflation happened, that dark matter is most likely made of WIMPS (see below), and, lately, that a dark energy pervades the cosmos.

Steady State — A rival theory to the Big Bang, championed most famously by the late Fred Hoyle. It says that the universe has always looked as it does today, and that as galaxies receded from each other, new matter appears that will eventually form new galaxies. Dropped by most astrophysicists like a hot potato after the detection of the CMB in 1965.

ylem — George Gamow's term for the hot cloud of matter that started the universe.

WIMP — Weakly Interacting Massive Particle, the leading candidate to explain dark matter, which presumably experiences only two of the four forces of nature: gravity and the weak interaction. It has never been detected, though physicists are still trying.

Acknowledgments

This book would have been impossible without the help of a group of astrophysicists and engineers who were extraordinarily generous with their precious time, even as they were working furiously to complete the most important and exacting project in the history of cosmology. First among these was David Wilkinson, who was also fighting lymphoma, the disease that finally defeated him just months before MAP's landmark results were announced to the public. Dave, whose career in cosmological research began when the Big Bang model of the early universe was considered a somewhat outrageous notion, was a gentleman, a brilliant teacher, a statesman of science, and, as his colleague Lyman Page said on his passing, "the best example I've ever seen of how science should be done." I had the honor of co-teaching a course with him at Princeton, but while we were equals in name, working with him was something like trying to keep up at basketball with Michael Jordan. Despite his illness, Dave was always willing to talk about cosmology and the MAP satellite, as long as we didn't dwell too much on his own accomplishments, which he inevitably downplayed. I am honored to have known him.

I am also grateful to the other Dave, David Spergel, who patiently led me through the complexities of theoretical astrophysics, explaining precisely what MAP would be looking for and where and how its discoveries fit into our emerging understanding of the infancy of our universe; to Chuck Bennett, leader of the MAP team, who was equally generous and adept at explaining the almost equally mind-boggling details of what it takes to design and build a satellite, and to navigate through the minefields of NASA's selection and approval process; to Lyman Page, Norm Jarosik, Ed Wollack, Al Kogut, Gary Hinshaw, Ned Wright, and all of the other members of the MAP team who gave me their time and were equally patient. David Spergel and Chuck Bennett were also kind enough to read the manuscript in search of the mistakes that inevitably crept in. It goes without saying, but I must say anyway, that any errors of fact or interpretation that remain in this volume are to be assigned as follows: me, 100%; them, 0%.

My editors at TIME—Jim Kelly, Steve Koepp, and Phil Elmer-DeWitt—were generous as well, in allowing me the time to work on this project even though they had a magazine to get out every week. My colleagues in TIME's science section, the best team of journalists I can imagine working with, took up the slack without missing a beat.

My wife, Eileen, and my daugher, Hannah, took up plenty of slack as well, patiently enduring my disappearances into the study for extended periods of writing and rewriting and frustration and then more rewriting. Their love and support mean everything to me.

Finally, I thank Laura Richman, whose transcription skills saved me many many hours of agony; my sometime teaching partner Ed Turner, whose insightful suggestions have led to more than one writing project, including this book; my literary agent Cynthia Cannell, whose patience, encouragement, and indefatigable support I privately don't think I deserve (but don't tell her, please); my editor at Princeton University Press, Joe Wisnovsky, whose calm professionalism and expertise in both science and writing have made this project go more smoothly than I could have imagined; and the Alfred P. Sloan Foundation's Public Understanding of Science and Technology program, personified for me by Doron Weber, whose generous support made this project possible. To all of these, and to those I somehow overlooked, thank you.

Index